DARWIN'S ARGUMENT BY ANALOGY

In *On the Origin of Species* (1859), Charles Darwin put forward his theory of natural selection. Conventionally, Darwin's argument for this theory has been understood as based on an analogy with artificial selection. But there has been no consensus on how, exactly, this analogical argument is supposed to work – and some suspicion too that analogical arguments on the whole are embarrassingly weak. Drawing on new insights into the history of analogical argumentation from the ancient Greeks onward, as well as on in-depth studies of Darwin's public and private writings, this book offers an original perspective on Darwin's argument, restoring to view the intellectual traditions which Darwin took for granted in arguing as he did. From this perspective come new appreciations not only of Darwin's argument but of the metaphors based on it, the range of wider traditions the argument touched upon, and its legacies for science after the *Origin*.

ROGER M. WHITE is Honorary Research Fellow at the University of Leeds.

M.J.S. HODGE is Honorary Research Fellow at the University of Leeds.

GREGORY RADICK is Professor of History and Philosophy of Science at the University of Leeds.

The cover image shows a Victorian Newfoundland water dog – bred not to fetch dead or wounded ducks but to save brave fishermen from drowning. In Darwin's *Notebook E*, p. 63, probably dated first week of December 1838, he wrote: "Are the feet of water-dogs at all more webbed than those of other dogs. — if nature had had the picking she would make such a variety far more easily than man, — though man's practiced judgment even without time can do much. — (yet one cross, & the permanence of his breed is destroyed)." It is the earliest surviving text reflecting Darwin's argument by analogy, any earlier ones having been cut out of this notebook by Darwin and not found since.

DARWIN'S ARGUMENT BY ANALOGY

From Artificial to Natural Selection

ROGER M. WHITE

University of Leeds

M.J.S. HODGE

University of Leeds

GREGORY RADICK

University of Leeds

CAMBRIDGE
UNIVERSITY PRESS

Shaftesbury Road, Cambridge CB2 8EA, United Kingdom

One Liberty Plaza, 20th Floor, New York, NY 10006, USA

477 Williamstown Road, Port Melbourne, VIC 3207, Australia

314–321, 3rd Floor, Plot 3, Splendor Forum, Jasola District Centre, New Delhi – 110025, India

103 Penang Road, #05–06/07, Visioncrest Commercial, Singapore 238467

Cambridge University Press is part of Cambridge University Press & Assessment,
a department of the University of Cambridge.

We share the University's mission to contribute to society through the pursuit of
education, learning and research at the highest international levels of excellence.

www.cambridge.org
Information on this title: www.cambridge.org/9781108708524

DOI: 10.1017/9781108769518

© Roger M. White, M.J.S. Hodge and Gregory Radick 2021

First published 2021
First paperback edition 2023

A catalogue record for this publication is available from the British Library

Library of Congress Cataloging-in-Publication data
NAMES: Hodge, M. J. S. (Michael Jonathan Sessions), 1940– author. | Radick, Gregory, author.
TITLE: Darwin and the argument by analogy : from artificial to natural selection in the
'Origin of Species' / Jonathan Hodge, University of Leeds, Gregory Radick,
University of Leeds, Roger White, University of Leeds.
DESCRIPTION: Cambridge, UK ; New York, NY : Cambridge University Press, 2020. |
Includes bibliographical references and index.
IDENTIFIERS: LCCN 2020054756 (print) | LCCN 2020054757 (ebook) |
ISBN 9781108477284 (hardback) | ISBN 9781108708524 (paperback) |
ISBN 9781108769518 (epub)
SUBJECTS: LCSH: Darwin, Charles, 1809–1882. On the origin of species. |
Darwin, Charles, 1809–1882. On the origin of species–Critism, interpretation, etc. |
Breeding. | Natural selection.
CLASSIFICATION: LCC QH365.08 H63 2020 (print) | LCC QH365.08 (ebook) |
DDC 576.8/2–dc23
LC record available at https://lccn.loc.gov/2020054756
LC ebook record available at https://lccn.loc.gov/2020054757

ISBN   978-1-108-47728-4   Hardback
ISBN   978-1-108-70852-4   Paperback

Contents

v

Preface

What can the actions of stockbreeders, as they select the best individuals for breeding, teach us about how new species of wild animals and plants come into being? Charles Darwin raised this question in his famous, even notorious, *Origin of Species* (1859). Darwin's answer – his argument by analogy from artificial to natural selection – is the subject of our book. We aim to clarify what kind of argument it is, how it works, and why Darwin gave it such prominence. As we explain more fully in our Introduction, we believe that the argument becomes much more intelligible when set, contextually, in a story stretching from classical Greek mathematics to modern evolutionary genetics: a long story, and a broad one too, encompassing everything from Darwin's earliest notebook theorising on the births and deaths of species, to agrarian capitalism as a distinctive form of economic life, to shifting Western reflections on art–nature relations.

A lucky conjunction led to our collaboration. On retiring from full-time teaching duties at the University of Leeds, RW (a philosopher who had written on analogy and metaphor) and JH (a historian of science often writing on Darwin) were asked to share an unusually spacious office. They quickly found that they had a common interest in Darwin's selection analogy, with RW seeing the first four chapters of the *Origin* as a shining example of how an argument by analogy ought to be conducted, and JH concerned to establish the place of the argument in the 'one long argument' of the *Origin* overall. They soon decided to try out their ideas in a seminar. It became clear that, approaching the same text from different angles, they had arrived at essentially the same interpretation of the argument, even though it was an interpretation at odds with much in the secondary literature. After the seminar, GR, also at Leeds, urged them to collaborate in writing about it for publication. Moreover, as he was himself a Darwin specialist, who had co-edited with JH *The Cambridge Companion to Darwin*, it was plain that this team of three friends should take on the task. The initial plan was for a long article, but it rapidly became clear that the material was so rich and complex that it demanded book-length discussion. In the course of working together we have arrived at a consensus on virtually all of the most important issues in the understanding of Darwin's text. We have much enjoyed learning a lot from each other, and strongly recommend such interdisciplinary teamwork as a way of academic life.

We are very appreciative of the encouragement and advice given us by Hilary Gaskin, our editor at Cambridge University Press. We have extensive debts to many people who have generously helped us as we have revised our draft chapters over several years. It is a pleasure to have this chance to thank André Ariew, Alex Aylward, David Depew, Jeanne Fahnstocke, John Henry, Emily Herring, Tim Lewens, Xuansong Liu, Charles Pence, Evelleen Richards, Robert Richards, Michael Ruse, Neeraja Sankaran, Prue Shaw, Elliott Sober, Susan Sterrett, Jonathan Topham, John van Wyhe, Pete Wetherbee, Gabrielle White, Polly Winsor and all who participated in several seminars over the last decade or so, culminating in a day-long workshop on the final draft in February 2020 at the Leeds Arts and Humanities Research Institute. Additionally, GR is grateful to the School of Philosophy, Religion and History of Science for the semester's leave which enabled him to complete his work on the book. All of us are grateful to Charlotte Sleigh for her expert indexing, and to our families for their patient support.

Introduction

Charles Darwin's *Origin of Species* (1859) argues for two big ideas, both expressed metaphorically: the 'tree of life' and 'natural selection'. New species of animals and plants have descended from earlier ancestral species; and these lines of descent with divergent modifications have branched and re-branched, like the branches on a tree. If all these lines trace to one first, common ancestral species, then all life forms one tree. Natural selection has been the main cause of these divergent modifications. By selective breeding, humans make, in a domesticated species, varieties fitted for different ends: strong, heavy horses for ploughing, and light, fast ones for racing. In the wild, over eons, natural selective breeding, due to the struggle to survive and reproduce ('the struggle for existence'), works unlimited changes in branching lines of adaptive descents, from fish ancestors fitted for swimming to bird descendants fitted for flying and mammals for running.

Our book is about Darwin's idea of natural selection. He called it that to mark the relation between selection in the wild and selection on the farm, or 'artificial selection'. Understanding this big Darwinian idea requires understanding his thinking about the relation between artificial and natural selection. Traditionally one considers, as Darwin did, how natural selection could be *analogous* to artificial selection, and how his argument from selection on the farm to selection in the wild could be an *argument by analogy*. But there are two difficulties. First, there is no consensus about what is meant by saying that two things are 'analogous', with specialists writing on Darwin no more in agreement than other writers on arguments by analogy. Second, several recent commentators have taken the radical revisionist line that, for Darwin, the relation between artificial and natural selection has been misidentified as one of analogy. But, once again, there is no consensus among these revisionists as to what the relation is.

We hold that Darwin was indeed arguing by analogy on behalf of natural selection, and that his analogical argument conformed to the oldest, ancient Greek view of analogy: the view taken by Eudoxus and, following him, by Aristotle, who construed analogy as proportion, as repeated ratio, as relational comparison. What is new in this book is the first sustained interpretation of Darwin's selection analogy as belonging in this distinctive tradition in the structural and functional understanding of analogy. We conclude that, in arguing from artificial to natural selection, Darwin was doing analogy, and doing it Aristotle style; that this was a good thing for him to be doing; and that he did it very well.

By way of a brief introduction to analogy as proportion, consider three examples, moving rapidly from the mathematical to the causal, and from the unremarkable to the remarkable:

- 1 is to 2 as 5 is to x.
- Socks are related to feet as gloves are related to hands. Since socks warm feet, gloves, which cover hands as closely as socks cover feet, are correctly inferred to warm hands.
- Stockbreeders are causally related to their livestock as the struggle for existence is causally related to wild organisms. The causal relationships are, in other words, the same in kind. But since the stockbreeders' selective breeding (artificial selection) is much less discriminating, comprehensive and prolonged – and so less powerful – than selective breeding by the struggle for existence (natural selection), the causal relationships differ in degree. Where artificial selection, the weaker cause, can produce only new varieties within extant species, natural selection, the stronger cause, can be inferred to produce comparably greater effects: not merely new varieties but new species.

Familiarly enough, '1 is to 2 as 5 is to x' specifies a mathematical proportionality. If, as here, three of the four terms are given, then – shifting from analogy to argument by analogy – the fourth can be calculated from them. Not so, of course, with the gloves analogy, or with the struggle-for-existence analogy. In these examples, given any three terms, empirical inquiry is required to establish the fourth. Furthermore, the relations in these examples are not mathematical but causal relations. Concentrating on what concerns us here, artificial selection mediates between its causes – the stockbreeders' actions – and its effects, the changes wrought in domestic animals; while natural selection mediates between its very different causes – the struggles for existence – and its very similar effects, the changes wrought in wild animals. The four related terms are not quantities, but the analogy is four-term proportional; and so an argument from this causal analogy is an argument from proportionality such as Aristotle was the first to analyse and validate.

In what follows we hope to persuade readers that placing Darwin's analogical argument from artificial selection to natural selection in the context of this Aristotelian tradition illuminates not only Darwin's argument but a range of topics extending well beyond it. We must emphasise, however, that it is no part of our brief to suggest that Darwin structured his argument as he did because he read Aristotle, or any later writer on analogy as proportion. As far as possible, we trace how the young theorist came to construct his causal theorising in that way; but we have found no reason to think that he was following what was said by any logical or rhetorical authorities on Aristotelian analogy. We shall say more on this topic in our concluding chapter, but for now, a parallel may clarify this issue. Like many scientific theorists, Darwin often constructed arguments conforming to the logical form modus tollens, or denying the consequent: the form of argument where the falsity of a statement is inferred from the falsity of another consequent statement that it entails. But bringing what logicians have said about modus tollens over the millennia to the examination of a Victorian scientific thinker's argument does not require believing that they learned from a logic book about this way of arguing. And so likewise, in our view, for Darwin's constructing his excellent examples of Aristotelian analogies.

****

Although this book is meant to be read straight through, an initial, high-altitude pass over its contents most usefully begins in the middle, with a trio of chapters (4–6) on the *Origin of Species*. Darwin called the *Origin* 'one long argument', and these chapters clarify how the whole argument is conducted, how Darwin's analogical reasonings about natural and artificial selection support his argument, and how his various metaphors are grounded in those analogical reasonings. Chapter 4 aims to show that Darwin structured the *Origin* as he did, and placed his selection analogising as he did within that structure, in conformity with a now unfamiliar ideal for the conduct of a scientific argument: the vera causa, or 'true cause', ideal. On Darwin's understanding of this ideal, it demanded, first, that the cause of interest be shown to exist, on the basis of evidence which is independent of what one is trying to explain; second, that, again on independent evidence, this cause is powerful enough to produce the effects to be explained, so that they *could have been* effects of this cause; and third, that this cause has *actually* been responsible for bringing about those effects. For Darwin, the argument by analogy from artificial selection to natural selection served to meet the second and third demands, by providing grounds for believing that, whereas selection on the farm could produce only new varieties within existing species, selection in the wild could go far further and produce new species.

Chapter 5 narrows the focus, from the overall structure and strategy of the *Origin* to this argument for the greater causal efficacy of natural selection compared with artificial selection. To secure this conclusion Darwin has to put in a lot of not-easy-to-follow work which, we suggest, is most easily grasped by seeing the argument as proceeding in two stages. In the first stage Darwin gives reasons for thinking that the same relation which holds between the stockbreeder and new varieties on the farm also holds between the struggle for existence and new varieties in the wild. In the second stage he gives reasons for thinking that, although the effects of the struggle will be the same in kind as the effects of the stockbreeder, the former effects can nevertheless be different in degree, accumulating to the point where not merely new varieties but new species are formed.[1]

---

[1] For readers eager for a more rigorous version: It is important to distinguish between Darwin's analogy and his argument based on that analogy. In its simplest form, Darwin's analogy has this structure:

The struggle for existence (B) is causally related to organisms in the wild (D) as the stockbreeder (A) is to organisms on the farm (C).

This is a statement of the analogy and not an argument to or for the analogy, nor an argument by or from it. Here, B and A may be called 'analogous' because, for some C and D, B is to D as A is to C. The two causal relations – natural and artificial selection – are analogous because their respective causes are.

Turning now to Darwin's argument, let us first consider the general structure of the simplest form of an argument from or by such an analogy:

A is F.
B is analogous to A.
Being F is invariant under this analogy.
Therefore B is F.

This is a valid argument form, so the strength of any argument with this structure depends on how well-justified the premises are.

Assimilating Darwin's initial argument by analogy to this structure gives us the following:

A stockbreeder (A) selectively breeds his or her animals and plants so that new domestic varieties are produced (is F).

The struggle for existence (B) is related causally to wild animals and plants (D) as the stockbreeder (A) is to his or her animals on the farm (C).

The same selective causal relation produces the same effect (being F is invariant).

Therefore, the struggle for existence selects in ways resulting in the production of new wild varieties.

When Darwin dwells upon the contrast between the weaker causal power of the stockbreeder and the much stronger causal power of the struggle for existence, he occasionally helps himself to metaphorical language – 'Man can act only on external and visible characters: nature cares nothing for appearances, except in so far as they may be useful to any being', and so on. Chapter 6 provides an analysis of these and other metaphors in the first four chapters of the *Origin* with a view to exploring their argumentational functions. Attention to these metaphors, in tandem with the analogies which they express, can help deepen an appreciation both of the potentialities of the argument-as-proportion tradition and of Darwin's skill in exploiting those potentialities. As we will stress throughout, when it is *relations* that are being analogised, the pairs of items bearing those relations can be strikingly different from each other. Moreover, once an initial analogy is in place, it can suggest extensions, which in turn can suggest further extensions. Shakespeare was a virtuoso of metaphors underpinned by imaginatively extended relational analogies. But Shakespeare wrote plays and poems, not scientific arguments. What makes Darwin's metaphors remarkable – and even more virtuosic than Shakespeare's – is their disciplined fealty to the analogies that carry parts of Darwin's argument.

By way of preparation for these *Origin*-centred chapters, our opening chapters (1–3) set out long-run, medium-run and short-run background stories. The long-run story, in Chapter 1, starts with Pythagorean mathematics, and with work, ascribed to Eudoxus, on proportion. It then moves to Aristotle, who showed how analogy as proportion could be deployed in a wide variety of empirical contexts, and who completed the Greek founding of the tradition of analogical reasoning most pertinent to Darwin's argument practices. When Aristotle affirms, for example, that scales relate functionally to fish as feathers do to birds, modelling of similar triangles is still a pertinent precedent. But the mathematical limitations are transcended for all posterity. Moreover, Aristotle emphasises that analogies can support insightful, suggestive metaphors such as one from later classical times: if fins are to fish as wings to birds, and fins are to water as wings to air, then we may say metaphorically that fish fly in water and birds swim through air.

The medium-run story, in Chapter 2, concerns the century and a half before Darwin wrote the *Origin*. On the one hand, this Greek tradition was alive and well in Darwin's day. On the other hand, this tradition no longer had a monopoly on even elite understanding of analogical arguments, with consequences that have sown confusion ever since, down to our own day. In particular, it was in the later eighteenth century that the Scottish philosopher Thomas Reid introduced an account of such arguments based not on proportion but on similitude. Reidian analogy is similitude between known and inferred properties, whether relational or not. Saturn, Mars and other planets are known to resemble the Earth in orbiting and being lit by the Sun, in having their own moons and so on. Inferably, then, they probably resemble the Earth in being inhabited. According to Reidian analogy, if two or more objects are all known to have certain properties, they probably also share other properties that some of those objects are not known but may be inferred to have. The shadow of doubt that now falls over whether analogical arguments, Darwin's included, can ever be really strong arguments is largely of Reid's making.

These two initial chapters cover millennia, centuries and decades. With Chapter 3 the pace slows to years, months and days. Here we unfold the short-run background story to the analogical argument of the *Origin*, covering the quarter century from Darwin's earliest causal–analogical conjectures about species extinctions, in 1835, through his pre-1859 theorising about species origins. By mid-1838 Darwin, in his private notebooks, had been for months comparing and contrasting species being naturally formed in the wild with variety formation under domestication. In doing so he distinguished between natural domestic varieties formed in regional isolation as adaptations to natural local influences such as soil, climate and vegetation, and artificial domestic varieties that are often monstrous and made by such unnatural arts as selective breeding. Naturally enough, he compared species being naturally formed in the wild with natural variety formation in domestic species, and insisted that nothing like artificial selective breeding was going on in the wild and influencing natural species formation. His arriving at his selection analogy, near the end of 1838, entailed a direct reversal of this comparison and this contrast. So Darwin in no sense discovered species-making natural selection via analogical reasoning from variety-making artificial selection. The point bears emphasis, because so many popular and even scholarly histories do not appreciate it.

After Chapters 4–6 come two final chapters which put our analysis of Darwin's analogical argument and its prehistory to work in various ways. Chapter 7 tests our reconstruction against the views of four revisionist commentators on the argument. We conclude, unsurprisingly, that none of the revisionists' principal proposals are reconcilable with our own or preferable to them. But in showing why, in the light of Darwin's texts and contexts, these proposals are unacceptable, we take full advantage of the opportunities offered to explore a diverse set of subsidiary topics, from his use of imaginative conjectures in the *Origin* to the possibility that his distinction between artificial and natural selection encodes a distinctly Victorian vision of social hierarchy. Throughout we try to underscore the value of an awareness of the analogy-as-proportion tradition in interpreting Darwin's analogising, in the *Origin* and beyond.

Finally, in Chapter 8, we consider the bearing of our analysis on wider disagreements about and within Darwinian science. Once again we return to Aristotle, to emphasise that the Aristotelian character of Darwin's analogical argument in no way implies that Darwin's science was Aristotelian, and also because Aristotle provides a useful point of entry into complex questions about the relationship between 'art' and 'nature' in Western thought. Whether we consider the Aristotelian tradition on that topic, or the tradition associated with the Aristotelianism-rejecting Robert Boyle, or the Boyle-rejecting tradition begun by the Romantics, Darwin's analogical argument appears on inspection to be a poor fit for all of them. Taking seriously Darwin's taking seriously the breeders' art helps too, we suggest, when we ask about the relationship between his science and the capitalism of his time and place, which was far more agrarian than tends to be remembered. Turning from pre-Darwinian to post-Darwinian contexts, we look, later in the chapter, at how the analogy remained instructive for three major theorists in the Darwinian tradition: Francis Galton, Alfred Russel Wallace, and Sewall Wright.

It is no purpose of ours to insist that Darwin's analogical argument must remain scientifically important for all time. If we enable readers to understand more fully how Darwin understood the argument, and to appreciate how considerable was Darwin's skill in putting the argument as he did, that will be achievement enough. Nevertheless, so long as Wright's side of his famous debate with Ronald Fisher on natural selection attracts proponents, so long, we suggest, will Darwin's argument live.

*****

In our experience as readers, a book like ours benefits from an introduction which supplies not only a high-altitude overview but a fairly detailed inventory of the chapter contents, the better to help readers see the wood for the trees (to invoke another venerable analogy). We close this Introduction accordingly.

Chapter 1:   Analogy in Classical Greece

Analogy as proportion first played a decisive role in science in solving the problem presented to early Greek mathematics by incommensurable magnitudes. Pythagorean mathematics taught that the relative magnitude of any two lengths, A and B, could be commensurably specified by two whole numbers, m and n, such that, if A is extended to m times its length and B to n times its length, then the two extended lines will be equal. But pairs of lines were later found not meeting this specification; and Pythagoras's own triangle theorem – equating the square of the hypotenuse to the sum of the squares of the other two sides of a right-angled triangle – proved this possibility. The Pythagorean account of relative magnitude was duly replaced by an account, almost certainly due to Eudoxus, that covered both incommensurable and commensurable magnitudes. It did so by specifying when four magnitudes, A, B, C, D, are such that A divided by B equals C divided by D (A/B = C/D); and so, proportionally, when A is to B as C is to D (A:B::C:D).

Here A and B must be quantities of the same kind, distances travelled, say; but C and D could be of another kind, times taken perhaps. A four-term relation allowed comparison of quantities of different kinds; and Greek mathematics took the word ἀναλογία as the name for such a four-term relation. In the theory of similar triangles this form of reasoning provided a valid proof by analogy for the further properties two such triangles must share, by treating each as a model for the other. This Euclidean and Eudoxian geometry included an initial examination of analogical relations and modellings pertinent to all our chapters here.

Analogy as proportion, as, more literally, repeated ratio, was soon moved by Aristotle beyond its mathematical confinements to diverse unmathematical, empirical reasonings. While remaining committed to proportionality itself as essential to analogy, he freed it from the limitation that when A is to B as C is to D, A and B must be items of the same kind, and likewise for C and D even if A and C are unalike. With this limitation removed Aristotle can argue to and from fins being to fish as wings are to birds. He can formulate analogies where the two objects being compared are, as he says, remote. For objects close in character, direct comparisons will be appropriate, especially comparisons identifying shared properties; but for objects remote in character indirect relational comparisons will be more apt: scales being to fish as feathers are to birds, or fins being to water as wings are to air. Remote objects can be compared by identifying their relations to other remote objects, in later lingo to other 'relata'. There was a price, readily incurred, for this new Aristotelian freedom. With a four-term relation among quantities, knowing the values of any three allows, by what became called the rule of three, calculation of the fourth; but, if it is not known what is to fish as wings are to birds, these three known terms do not determine what this unknown fourth must be. Only empirical inquiry into fish structures and their functions can do so.

Such empirical relational comparisons play major roles in the comparative teleological anatomy of Aristotle's biological works. In these indirect comparisons, two animals as unlike as a bird and a fish can be models of each other.

Aristotle's biology was not called 'biology' and was not biology as Darwin's generation would know it. Aristotle's cosmology and his metaphysics, the foundations for his science of life, were no longer foundational for natural history and comparative anatomy more than two millennia on. But the legacy of his theory and practices of analogical comparisons endured. Darwin had to hand on HMS *Beagle* a brand-new little book – by the Oxonian John Duncan on *Analogies of Organised Beings* – a book acknowledging Edward Copleston and his former tutee Richard Whately as mentors who had enlightened the author about analogy, as proportional, relational likening, as taught by their own mentor Aristotle. Darwin's copy has no annotations so he is unlikely to have read it carefully and profitably.

Chapter 2: Analogy in the Background to the *Origin*

Mediaeval philosophers of all three leading Abrahamic faiths deployed Aristotle's teachings in their novel analogical comparisons of talk about God and about his creatures. The scholastic authors of the high middle ages, in their precision and sophistication, emulated their master, and in doing so gave 'analogy' new uses and meanings. Aristotle in presenting his account of analogy had talked of words 'being said in many ways', as in saying 'A is F' and 'B is F' when A and B have no common intrinsic property. Such cases include not only proportional analogies, but also cases where, for example, some diet for cows is said to be 'healthy', because it causes cows to be 'healthy' in what is today often called the 'focal meaning' of this word. Perhaps because they misread Aristotle, the school men called all these cases instances of 'analogy', while retaining the contrast between analogy and simple similitude. With their preoccupation with analogy's implications for such ontological and linguistic questions, they had little interest in argument by analogy. So, in preparing historiographically for our chapters on Darwin's *Origin,* we write only very briefly about what Aristotle's mediaeval followers did with his legacy as analyst of analogy.

The precision and sophistication of the scholastics was not emulated by Renaissance and Enlightenment authors in the sixteenth and seventeenth centuries, in their revivals of Epicurean, Stoic and Platonic alternatives to Aristotle's legacies for philosophy and for the sciences. Today's historical dictionaries for vernacular European languages, like the encyclopaedias from those centuries, confirm that 'analogy' and its cognates became used in diverse and casual ways, acquiring many uncoordinated meanings with little in common except some association with 'similarity'. These undisciplined discursive habits extended into the early eighteenth-century decades, when the battle between the ancients and the moderns turned in favour of authors declining deference to Greek and Roman antiquity. Within the norms of his time, Joseph Butler countered deism, in his 1736 book on the *Analogy of Religion, Natural and Revealed, to the Constitution and Course of Nature,* with no explication for the leading term in his title.

As we have already noted, an enduring alternative to the Aristotelian view of analogy as relational comparison traces to Thomas Reid, over two decades before Darwin's birth, and is still prominent today. And by the beginning of Darwin's century, there were three influential clarifications of analogical reasoning. Kant in Germany and Copleston in England independently returned to Aristotelian analogy as proportion, making no concession to the Scotsman Reid's recent version of analogy as similitude. Kant drew mainly on Aristotle himself. With his concern to demarcate cognitive roles for reason and experience, Kant emphasised the differences between analogies constituting a priori mathematical knowledge, and those contributing to empirical knowledge a posteriori; and so he dwelled especially on analogies, prominent in the natural sciences, asserting sameness of causal relations and supporting inferences from the known to the unknown consequences of those causal relations.

In England, at Oxford, Copleston saw himself as in descent from William King, an Irish Anglican bishop who, along with Peter Browne and the more famous George Berkeley, had discussed, early in the eighteenth century, the implications of analogy as proportionality for venerable questions concerning human knowledge of God. In the late 1820s, Whately, another Oxonian affirming his debts to King and soon to be an Anglican Archbishop in Ireland, gave analogy as proportion a place in both his logic and his rhetoric texts, just as Aristotle had. Well into Darwin's adult life, Whately stayed resolutely committed to the Aristotelian understanding of analogy; while John Stuart Mill carefully distinguished Aristotelian and Reidian arguments by analogy, regarding both as legitimate, although very different, forms of argument.

Darwin was sent to Cambridge by a cynical unbelieving father who thought a career as a Church of England priest would suit his son better than an earlier curricular choice, medicine, had done. Convinced that he was himself a sufficiently believing Christian this student son studied carefully two books by Archdeacon William Paley, one on revealed and the other on natural theology. Later, when reflecting on these two theologies, he was indebted to Paley's typically eighteenth-century evidentialist view of religious belief. On analogy, Paley had no sustained stance to teach. His arguments to divine intelligent design in nature do not exemplify any analogical forms of reasoning, whether Aristotelian or Reidian. Nor were Darwin's analogical reasoning practices instructed by any reading in any writings on analogy in general; but our readings of Darwin can be instructed by those writings even if his writings were not.

Chapter 3: Darwin's Analogical Theorising before the *Origin*

Darwin arrived at his selection analogy very late in 1838. Nearly four years earlier he had begun his first sustained proportional, analogical reasoning on behalf of a causal–explanatory theory. This theorising, about species mortality and so species extinctions, is the single most instructive precedent for Darwin's analogical reasoning about natural selection.

Darwin opened reflections on species extinctions, in February 1835 when still in South America, by admitting of an 'analogy' that it was a 'false one', but then adding a 'but' and defending it. With apple trees, artificial grafting could vastly extend the life of one bud but, horticultural lore agreed, only extend it limitedly as if the successive trees propagated by repeated grafted cuttings were all parts of one tree. So, Darwin argued, the lifetime of a mastodon species could be extended vastly but limitedly by natural sexual reproduction.

In spring 1837 this analogy of Darwin's, although not claimed to be 'close', is no longer deemed false. He has now developed an explicit foundation for it: all generation, sexual and asexual, natural and artificial, extends life by reiterated divisions of individuals beginning with an initial individual containing a limited, finite quantity of extendible life. So, natural sexual generation is to the extending and ending of an animal species as artificial grafting is to the extending and ending of a grafted plant succession. The common causal relation of limited generational division has a common causal consequence: the limitation of extended life.

Darwin's main mentoring as a theorist of generation was at Edinburgh before going to Cambridge. On the voyage, he drew on Lyell's respectful review and rejection of inherent species mortality theorising in explaining species extinctions. There is no reason to think that Darwin's development of his analogy drew directly on any general analyses of analogical reasoning. And it was to be likewise with his first analogical reasonings about natural selection.

Reflecting on Malthus in late September 1838, Darwin took the high potential rates of animal population increase to vindicate Lyell's view of species extinctions as due, not to any inherent species mortality, but to terminal competitive imbalances caused by very slight changes in local conditions. So much for losing species. Darwin says that the 'final cause' of population pressure, its good consequence, presumptively Divinely intended, is to 'sort out', discriminatingly retain, in winning species, adaptive structures fitted to changing conditions. But more than two months will pass before he likens this sorting to artificial selection, and drops his longstanding contrast between species formation in the wild and the making of domesticated varieties by selective breeding.

Around 30 November, Darwin declared, in unusually triumphant mode, that three numbered principles will account for everything: a principle of heredity, of variation and of superfecundity. These three principles do indeed constitute what he will eventually, in the early 1840s, call his theory of 'natural selection'; but it cannot be called this yet because there is still no selection analogy. Within days comes the first talk of nature's picking or selection, with immediate emphasis on this selection by nature being far more powerful than artificial selection. By mid-March 1839, the reversal concerning varieties is explicitly decisive for Darwin's plans for public argument: some domestic varieties are formed by natural adaptations to local conditions, but greyhounds, pouter pigeons and others are made by the arts of crossing and selection. Hence, next, this question: does nature have 'any process analogous'? And hence Darwin's resolving to answer by introducing his theory, later texts showing that he would here invoke superfecundity as causing the struggle for existence and its natural selective breeding consequences.

Darwin had not reasoned thus: domestic varieties are made by selective breeding; wild varieties are like domestic varieties; like effects have like causes; so wild varieties are made by something like selective breeding in the wild. Had he done so, the existence of selection in the wild would have been inferred from its existence on the farm. No, Darwin reasoned that, owing to the struggle for survival in the wild, sorting exists in the wild and causes adaptive change; and owing to stockbreeding practices there is selective breeding on farms causing diversification of domestic varieties; and (on subsequent reflection), the discriminating sorting effects of the wild struggle are seen to resemble the effects of the breeders' art. Hence it is not true (as he'd previously thought) that there is nothing like selection going on in the wild. Here the existence of selection in the wild is inferred not from the existence of selection on the farm, but from the consequences of the existence of the struggle for survival in the wild; the struggle, in its selective actions and their consequences, is then causally to wild animals as the stockbreeder is causally to farm animals. Same causal relation and same consequences. Thus did Darwin arrive at this four-term causal–relational comparison, this analogical proportion.

The construction of the theory of natural selection and its analogical articulation had resulted from a complex series of steps in nearly half a year from September 1838 to March 1839; steps that, with allowable hindsight, can be read as eventually, gradually bringing him to this result. In 1842 Darwin wrote out a brief rough sketch of his theorising. It would be the textual ancestor to the book-length draft essay of 1844; and so too to the very big book he was calling *Natural Selection* and composing, in 1858, when Wallace's letter and essay prompted him to prepare the *Origin* as an abstract. In these years after 1839 and before 1859, Darwin had various new unpublished thoughts about natural selection. But the theory was never modified, then, or indeed after 1859, in any ways requiring modifications to the selection analogy.

Chapter 4: The 'One Long Argument' of the *Origin*

The *Origin's* first edition has an introduction on how Darwin came to write it and what its fourteen chapters are about. Opening the final chapter he calls the book 'one long argument', and then recalls the argument's main elements, before closing with their implications for future science.

Within the thirteen chapters before the final one, the relationship between the initial four and last five is decisive for the structuring of the long argument; for the argument makes three evidential cases for natural selection: the case for the existence of this cause, and the case for its power to form and adaptively diversify species (I–IV), and the case for its having been responsible for the forming of extant and extinct species (IX–XIII). Existence, adequacy and responsibility. It is, it could and it did. In that order, because only an existing cause could be adequate, and only an existing and adequate cause could have been responsible. Hence, the initial four chapters marshal evidence for existence and adequacy, and the last five for responsibility. Hence, in the intervening four chapters (V–VIII), one chapter complements the book's opening two on variation on the farm and in the wild, and three then counter objections to the adequacy case made in chapter four. Chapter IX is sometimes associated by Darwin with this sequence of intervening chapters. But it properly belongs with the later sequence in countering in advance objections to the responsibility case.

The *Origin* offers scant explicit guidance on how the long argument is conducted but the unpublished drafts help. For the 1842 *Sketch* and 1844 *Essay* are conformed more discernibly to the vera causa (true cause) evidential ideal. The ideal had had its canonical formulation in Thomas Reid's brief explication of a Newtonian methodological dictum, but Reid did not associate it with his view of analogy. Among the scientists most respected by Darwin, the ideal was upheld by Lyell and by Lyell's friend and geological ally, the physicist and astronomer John Herschel; and it was later rejected by Lyell's critic and Herschel's friend, the polymath Whewell. The agreement of the two Reidians, Lyell and Herschel, undoubtedly influenced the young Darwin but precisely when and how is not known.

A traditional requirement for any causal–explanatory hypothesis was that the cause it invokes be evidenced as capable of producing the kinds and sizes of effects it is to explain. Further, a good hypothesis explains many different facts about those effects. Such explanatory virtue is evidence for the responsibility of the cause for the effects and so for its existence; but, because this evidence for its existence is not independent of these facts, the hypothesis is deemed to be conjectural, speculative and the cause hypothetical. By contrast a vera causa is a cause that has its existence also evidenced independently by facts other than those it is to explain. So, to show that some causal–explanatory theory is no mere hypothesis but a vera causa theory and hence inductive, not conjectural, the two requirements met by any good hypothesis must be met and a third also: independent evidence that the cause is real, true, exists and is known. Darwin's theory is that natural selection with arboriform descents is responsible: it did it. This is the punchline for the whole book, the conclusion of its long argument. By arguing for the existence and adequacy of this cause, the first four chapters enable those later five (IX–XIII) to argue for this conclusion, and so for this theory.

The selection analogy has no role in the case for the existence of hereditary variation, nor in the case for the existence of natural selection. The opening chapter gives evidence for the existence of hereditary variation under domestication, and the second chapter argues from this to the existence of such variation in the wild. The causes of all hereditary variation are the same: changes in conditions, in soil, climate, food and so on. Such changes are known to be effective under domestication. Geology, Lyell's geology, shows such changes going on everywhere and always in untamed nature, and they can be inferred to have similar effects in causing hereditary variation albeit less abundantly than on farms and in gardens. The existence of deliberate selective breeding on farms and in gardens has been widespread in recent decades. But there is no inferring from this that there exists selective breeding in nature. As argued in the third and fourth chapters, the existence of natural selection can be inferred from excessive animal and plant fertility, and the consequent struggle to survive and reproduce in the wild. For in this struggle there will be consistent, persistent and comprehensive discrimination among hereditary variants, favouring and so preserving advantageous variants in successive generations.

In this discrimination the struggle is acting as a stockbreeder does in forming and diversifying animal varieties. But with two differences: whereas the stockbreeder forms varieties fitted to their uses or fancies, the struggle forms varieties adapted to the animals' and plants' own ends of survival and reproduction; and, second, while these effects of the struggle are the same in kind as the effects of the stockbreeder's artificial selection, the effects wrought in the wild can be vastly greater in degree, including the transforming of wild varieties into new species; for varieties and species differ only in degree not in kind, varieties being incipient species, and species well-marked varieties. So, over eons, selection in the wild can cause the adaptive diversification of the entire tree of life. The selection analogy here supports an argument a fortiori, from the greater strength of natural compared with artificial selection. The fourth chapter, and the case for the adequacy of natural selection, can then end with the principle of divergence. Since structural and functional specialisation is usually advantageous in life's struggle, over long ages natural selection causes, reliably if not invariably, branching and re-branching structural and functional divergences in favouring diverse adaptive specialisations, with more specialised species winning out over the less specialised which become extinct. Specialisation being a criterion for progress, this adaptive tendency is progressive.

These – the conclusions in Darwin's fourth chapter, about the powers and tendencies of natural selection in divergent branching modifications of common ancestral inheritances – are conclusions about the tree of life and natural selection, and so about the implications of each big idea for the articulation of the other. First fully integrated in this most crucial chapter, the two big ideas remain so for the rest of the book: being defended in the middle four chapters, and deployed in the five following them in formulating explanations for many kinds of geological, geographical and morphological facts about extinct and extant species. The conclusions of the fourth chapter are deployed in these explanatory tasks because they have established what consequences branching natural selection, the cause, can have, and therefore what branching natural selection, the theory, can explain and how. The special conclusions of the arguments from the selection analogy are then decisive for the long argument of the whole book, a long argument conformed to the Reidian vera causa evidential ideal, not however to Reidian analogy as similitude, but to Aristotelian analogy as proportion.

Chapter 5: An Analysis of Darwin's Argument by Analogy

In modern times, the reputation of analogy has suffered from associations with unfashionable mediaeval scholastic theological apologetics, and from confusion about how analogical reasonings are formally structured and so logically assessed. With Mill, if not with Copleston and Whately, those old associations fade away; but there remains today a need for elementary formal precision.

With Aristotelian analogy as proportion the argument from a four-term relational comparison is an argument from one or more premises about an analogical model, to one or more conclusions about an analogical target. Suppose a situation or object or state of affairs, M, is an analogical model of target situation T. If M is F and being F is invariant under the analogy, and so inferably transferable from M to T, then T is F. An oven baking dough is a model for the sun shining on mud. The oven is F, is drying and so hardening what it bakes. Drying with hardening (F) is inferably transferable from model to target, therefore the sun is F, is drying and hardening what it shines on. Here A (the oven) having relation R (heating) to B (the dough) is a model for the target, C (the sun) having the same relation R to D (the mud). The model is what is learned from and the target is what is learned about. The modelling scheme can be expanded by adding that the model M is G and that M's being G suffices for M's being F; and adding that T is G. Then it follows that T is F. The oven being hot (G) suffices for the drying with hardening (F); the sun is hot (is G); so it dries and hardens (F) the mud. This is valid analogy, and suggestive justification for saying that the sun bakes the mud, and so for talking metaphorically as if a culinary art is practised by solar nature. Reasoning from the oven's being hot is meeting Mill's requirement that any reasoning from a proportional analogy should be supported by a material circumstance; by a circumstance relevant to the model having what is invariant under the inferential transfer. Here there is causal relevance; for this circumstance, the heating, is common to model and target, and has as a causal consequence the invariant, transferable feature of the model, the drying with hardening.

As with many argument forms, so with reasoning from analogical modelling: simple schemata and homely examples can make the whole business look commonsensical, but hardly capable of instructive scientific sophistication. And there are indeed only two simply specifiable requirements for validity in an argument from a four-term relational comparison asserting that A is related to B as C is to D. For to infer validly, from this comparison, that D shares some feature known to be possessed by B, it is necessary that the common relation does truly hold in both cases; and that, in conformity with the material circumstance requirement, in both it does truly have the same consequences. Same relation and same consequences of same relation: those are the essentials. Told by Darwin that the stockbreeder is related to the animals on the farm as the struggle for existence is to those in the wild, his readers have to ask if these relations and their consequences are truly the same, or at least sufficiently relevantly similar to support the conclusions Darwin is drawing from the four-term comparison.

Answering these questions is often far from easy, thanks to Darwin's concern with detailed, actual exemplifications and with pre-emptive counterings of objections to his premises and conclusions, all in prose rarely deft and lucid. Many exegetical difficulties diminish if we separate, as Darwin does not, two stages in his reasoning from premises about his analogical model, artificial selection on the farm, to conclusions about his analogical target, natural selection in the wild. The first stage argues that just as artificial selection forms new domestic varieties, so natural selection can form new wild varieties. The second stage, building on the first, argues that natural selection, unlike artificial selection, can form and diversify without limit new species, and so species of new genera, families, classes and so on. In both stages the material circumstance requirement is met by establishing that, in both artificial and natural selection, there is consistent and persistent favouring of some hereditary variants over others throughout successive generations. So, in both stages, artificial and natural selection are being compared qualitatively as causal processes of the same kind. But in the second stage they are also being contrasted quantitatively as causal processes differing vastly in degree. A brand of bitter ale used to be advertised as like beer but more so. Darwin's natural selection is like artificial selective breeding but more so.

The greater power of natural selection is manifest when considering the more precise, comprehensive and prolonged discriminations among hereditary variants effected by the struggle for existence: the struggle due to the tendency of reproductive multiplication to produce more offspring than can be sustained by the limited resources, of food especially, that sustained the parents' generation. Arguments from this greater power are arguments a fortiori. Darwin's a fortiori arguments are not arguments making stronger the reason for accepting some conclusion; they are arguments from the greater strength of one cause compared with another, and so to the greater extent of one lot of effects over another.

A fundamental feature of analogical proportion, alternation, is exploited here. 10 is to 5 as 2 is to 1 can be validly alternated to 10 is to 2 as 5 is to 1. Likewise, if artificial selection is to the making of domestic varieties as natural selection is to the forming of wild varieties, then, by alternation, what the stockbreeding community does is to nature's workings as the making of domestic varieties is to the forming of wild ones. With this alternation, a vast scaling up of farm life can be a model for the alternated analogy's target: the whole living world.

To summarise, natural selection is to the effects of natural selection as artificial selection is to the effects of artificial selection. Therefore, natural selection is to artificial selection as the effects of natural selection are to the effects of artificial selection – the more powerful the cause, the more powerful the effect, and the achievements of natural selection may far surpass those of artificial selection.

Chapter 6:   Darwin's Use of Metaphor in the *Origin*

If analogy needs rescuing from unfashionable associations, metaphor needs rescuing from fashionable ones, especially from recent if now passé vogues for Nietzschean one liners about truth being a mobile army of metaphors. In understanding Darwin's uses of metaphor, rescuing can begin with Aristotle's views on what analogy and metaphor can do for each another.

If real, analogical relations are in the world. If meaningful, metaphors are effective practices within language. Metaphors can suggest analogies; and analogies can support metaphors; and, as Aristotle taught, proportional analogies are especially good at doing so.

An analogy can support more than one metaphor. By elaborating the relational comparison the analogy is making, it can support a metaphor's extension in further successor metaphors. An initial comparison of Shakespeare's Antony with the Sun can support talking metaphorically of his death as a sunset. With additional analogical comparisons between this death and a sunset, further reiterated supportings are possible. With artificial selection as his analogical model for natural selection, Darwin can talk metaphorically about nature's discriminating among animals in the wild, as if he were talking about a farmer discriminating among his livestock.

In supporting metaphors with analogy Darwin, man of science, had to be constrained in two ways not expected of the poet Shakespeare composing his play. Darwin's metaphors had to be amenable to literal, if cumbersome, paraphrase informed by the supporting analogy; and they had to contribute to the argument of his book. Many of the Bard's metaphors resist literal paraphrase; and while his play may present arguments, he has not composed it as one long argument for a causal–explanatory theory.

Throughout the *Origin*, Darwin is deploying metaphors as convenient and vivid shorthand; and he is elaborating metaphors in making extended comparisons between life in the wild and on the farm, life in the short run of the present and long run of the past.

His metaphors contribute also to concept formation. The physicists of his day were likewise giving words new meanings by giving them new metaphorical uses to meet needs for new concepts. Traditionally work was what a human or a horse did in tilling a field. But the word 'work' could be used metaphorically of what a machine did in a factory or on a railroad track; and here the work of any machine in a specified time could be quantified as equivalent to the lifting of a standard weight through a standard height; and this physicists' concept of work could feature quite abstractly in developing thermodynamical theory. The word would keep its workaday uses and meaning, while theorists of thermodynamics would no longer align their term and its meaning, their concept, with toiling and tilling talk.

Darwin's concepts were not as quantitative and abstract as the physicists'. His selection and struggle talk stayed closer to pub and street talk than theirs did. His selection and his struggle metaphors contributed to the formation of his concepts of natural selection and of the struggle for existence, in ways making their phrasings open to varied interpretations and persuasive uses leading sometimes to confusion and dispute among his readers. His own uses of the phrase 'natural selection' may take both words metaphorically rather than literally. But sometimes the first word is taken literally and the second metaphorically, or indeed vice versa, while again both words may be taken literally. In using metaphors Darwin was obviously not an innovator, but his uses for metaphors were not casual and conversational. Many of his most sustained metaphors are in his book because they are prompting suggestively his deployments of his analogical modelling which, in return, is grounding and supporting those metaphors.

Chapter 7:   Rebuttals of the Revisionists

Until recently there was consensus that Darwin's *Origin* argues analogically for the theory of natural selection, although there was no agreement about how the analogical reasoning goes. Lately this consensus has been challenged by several revisionists who, while disagreeing among themselves as detailed earlier, all see analogy contributing little to Darwin's long argument. On our reading of the *Origin's* long argument, Darwin is integrating throughout the Reidian vera causa ideal and the un-Reidian, Aristotelian norm of analogy as proportion. The four revisionist views rule out any such interpretation, which is why we have weighed them carefully, before concluding that they are not established by textual evidence.

James Lennox accepts that Darwin conformed his theorising to the vera causa ideal, but proposes that he met the ideal's adequacy requirement not through analogical comparing of artificial and natural selection, but through imaginary illustrations of natural selection called by Lennox 'Darwinian thought experiments'. In Lennox's analysis, rejection of the analogical interpretation is implicit. Richard Richards, Peter Gildenhuys and D. Graham Burnett have rejected it explicitly but without agreeing on where it goes wrong and what should replace it. Like Lennox, Richards sees Darwin as meeting the adequacy requirement by appealing not to any analogy between artificial and natural selection but to something else: not thought experiments, however, but real experiments, accidentally conducted when selectively bred domestic animals have gone feral. For Richards, Darwin made use of the selection analogy only as a heuristic guide to how natural selection works. Gildenhuys, even more radically, thinks the vera causa ideal irrelevant and denies that Darwin recognised analogous forms of selection. He presents Darwin as using a selection analogy only in establishing a completely general account of selection wherever it is found, and however much its various forms seem to differ. Burnett is no less radical in a different direction, stressing Darwin's appeals to diverse forms of selection. Darwin's use of them to fill the gap between artificial and natural selection results in a spectrum so continuous, Burnett says, that it subverts any analogical argument. Far from this subversion weakening Darwin's case for natural selection, however, it is, in Burnett's view, this very collapsing of analogy into identity that gives the long argument its strength.

Sometimes exegetical disagreement has to be unequivocally articulated if it is to be honestly and respectfully expressed. Our disagreements with these four revisionist views bar any compromise or assimilation. For, like all earlier interpretations of Darwin's reasoning in the *Origin*, none of the recent revisionist proposals takes any sustained account of the Aristotelian tradition of analogy as proportion, whereas our reading of Darwin brings that tradition to the interpretation of the overall structure and the detailed content of Darwin's long argument.

Chapter 8: Wider Issues Concerning Darwinian Science

Placing Darwin in the long run of Western thought concerning relations between art and nature requires at least two contrasts. The art of selective breeding is not like the art of a doctor doctoring himself; but, for Aristotle, thinking of an acorn artfully turning itself into an oak tree, such a doctor is what nature is like. Nor is selective breeding like the art of a seventeenth-century Boylean or nineteenth-century Paleyan watch maker, who makes a small mass of passive, inert metallic matter into an intricate working machine that this matter could never turn itself into. The selective breeder's mindful art works with the active powers of reproduction and hereditary variation; so too does the mindless, artless but artlike, natural struggle for existence as the selector in the wild.

Stockbreeders make and improve breeds, varieties, by sustained selecting among individuals, selecting which eventually results in better individual living machines later. But this improving of the individuals is not done by going to work on each individual as an individual doctor doctors himself, or as a watchmaker makes one watch at a time, but by going to work on a flock, a herd, or an orchard over generations.

So, there is no significant likeness here to the inventing and improving of industrial machines for machino-facturing capitalism; but there are manifest alignments with agrarian capitalism's profitable improvements of animal livestock and crop plants by selective breeding. Nature's selector, the struggle, is a Malthusian consequence of reproductive fertility outrunning food resources; and land use for food production was the link between Malthus on population and Malthus as defender, against the machino-facturing interests, of Corn Laws, and as defender in economic theory of earlier French physiocratic privileging of agrarian capital over all other capital. So, Darwin's selection analogy is not descended from a venerable tough-minded Cartesian mechanist metaphysics newly inspired by a youthful English machinist industrial capitalism. Is it then to be read as grounded in a tender-minded, pantheist, even pananimist view of nature owing to Darwin's debts to Humboldt and so to expressivist, aestheticist, idealist German Romanticism? Hardly. Darwin in the 1830s is no atheist but no pantheist either. Nor, for Darwin, is all nature ensouled. Again, German romantics, Schelling, the brothers Schlegel and their like, privileged art over science and religion in their views of man and of nature, but not art as craft but art as poetic, creative expression of human selves; art as Beethoven or Schiller not art as Bakewell the Leicestershire stockbreeder known as the man who invented sheep.

Placing Darwin in a tradition of analogy as proportionality requires seeing how he could be in this tradition, and yet far from perpetuating any Greek traditions in metaphysics and cosmology, whether Epicurean, Platonic, Stoic, or indeed Aristotelian. Turning from the long to the short run, it means seeing how he could be – like his main mentors as a scientific theorist, Lyell and Grant – predominantly Scottish and French in his scientific culture and therefore only very slightly Germanic and far from predominantly English. The Aristotelian tradition of analogical–proportional reasoning had no inherent connection with one view of man and nature rather than another. The content of Darwin's selection analogy has its historical sources, but that content is no more owed to the German Kant or Scottish Mill than it is to the Greek Aristotle, even though the analogy's structure and function does align him with their shared proportionality view of analogy.

Aristotle on analogy, Reid on true causes and Darwin's agrarian modelling for life in the wild are directly relevant to evolutionary biology in our time. What is arguably still the most influential theory of evolutionary genetics, Sewall Wright's shifting balance theory, was developed and defended by him as modelling evolution on artificial breeding practices, including especially the carefully recorded breeding practices of shorthorn cattle farmers in north-eastern England in Darwin's century. Wright and fellow founders of mathematical population genetics, Fisher and Haldane, brought statistical theory to their modellings of evolutionary processes, but statistics as analyses of causal processes – selection, inbreeding, mutation, random drift and the rest – not as mathematical replacements for causal analyses. Darwin and his recent scientific descendants, for all that they are obviously modern in attitudes and achievements, nevertheless draw in their work on traditions going back centuries, even millennia.

CHAPTER I

Analogy in Classical Greece

The aim of this book is to analyse and defend the claim that the first four chapters of Darwin's *Origin* constitute an argument by analogy from artificial selection to natural selection, situating that argument in Darwin's thought as a whole: just as human beings by their selective practices in domestic settings can make new varieties of plants and animals, so the struggle for existence in the wild can make new varieties and even new species of plants and animals. This claim has been frequently made, but also latterly contested. However, both the defenders and the opponents rarely spell out in detail what the argument is supposed to be, and, insofar as they do so, usually work with an inappropriate account of what an argument by analogy is thought to be.

Therefore, before turning to Darwin himself, we need, in this chapter and the next, to examine the idea of an argument by analogy. We begin in classical Greece where the concept of analogy was introduced, before turning in the next chapter to the emergence of a completely different conception of analogy, and with it a completely different account of argument by analogy. We shall argue that although the later account has become the most popular understanding of 'argument by analogy', it is the classical account which is the appropriate one to account for the text of the *Origin*.

The point is that the word 'analogy' has historically been understood in two quite different ways. The word was initially introduced in Pythagorean mathematics ('ἀναλογια') and then extended into the empirical domain, above all by Aristotle. Here, the word always designated a proportionality ('A is to B as C is to D'), and the interest was in the rich variety of uses to which appeals to analogies of this kind could be put, as against simple similarities ('A and B share some intrinsic properties'), whose uses were very limited. The contrast between analogy and simple similarity was always observed and insisted upon. However, beginning in the seventeenth century, in large part as a reaction against mediaeval scholasticism, this contrast was ignored – or, at most, it was noted that this distinction was important but only in mathematics. Elsewhere 'analogy' was treated as a near synonym for 'similarity'. Insofar as the two were distinguished, it was not in accord with Greek usage, but talking of 'analogy' was treated as most appropriate when A and B shared several similarities.

Corresponding to these two different ways of understanding the word 'analogy' were two completely different accounts of what constituted an argument by analogy. Without at this stage analysing them in detail, we may look at the following two supposed uses of 'argument by analogy'. Consider first the following 'anti-democratic' argument ascribed by Aristotle to Socrates,[1] where Socrates is protesting against the use of a form of lottery in the appointment of certain offices of state (a procedure deemed 'democratic' because every citizen had the same chance of holding office):

> We ought not to choose our magistrates by lot, since this would be like choosing the athletes to represent us at the Olympic Games by lot rather than by their skill at athletics, or like sailors choosing their helmsman by lot, rather than one with the relevant knowledge.

When we realise why it would be absurd to choose athletes or helmsman by lot, since there are skills vital to being a successful athlete or helmsman, we see that by analogy it is absurd to choose magistrates by lot. What we have here is an argument by analogy that is valid, given the tacit premise that there is a range of skills necessary to carry out the tasks of a magistrate successfully.

Contrast this with:

> This berry shares a large number of characteristics with a berry known to be poisonous. Therefore it is probable that it is also poisonous. The more characteristics it shares, the more probable it becomes that it is poisonous.

Both of these arguments clearly need tightening up: in the first case, we need to show that the analogy holds in all relevant respects, and in the second, we need, for instance, to find a way to exclude characteristics that are irrelevant to the point at issue. But even after tightening up, what we have here are two radically different arguments. The first argument can be developed into a fully valid argument, but the second, though not worthless – it is obviously sensible to avoid the berry – is at best a probabilistic argument, beset with difficulties in attempting, for example, to quantify

---

[1] *Rhet.*, II. 1393a 23–1393b 7.

the probabilities involved. Because of this, it is clearly crucial when we come to examine Darwin's presentation of his argument by analogy in the first four chapters of the *Origin* to be clear which of these two patterns of argument is involved. This is particularly true since the vast majority of commentators, both those claiming that these chapters do constitute an argument by analogy and those who contest this, have assumed that there is thereby meant an argument of the second sort, whereas it is the central contention of this book that a careful examination of Darwin's text shows him as presenting a near perfect argument of the first sort.

Therefore, before turning to our main subject – an exegesis of Darwin's use of the analogy – we look further at the two different forms of argument. In this chapter, we look at the introduction of analogy as proportionality in classical Greece, including the prototype of the first form of argument by analogy, the development in Euclid VI of the theory of similar triangles, and hence, by extension, the simplest of all analogical models – the scale model. Then in the next chapter, we look at the centuries immediately prior to the *Origin*, where we find two things: the emergence and eventual great popularity of the second form of argument, while alongside this we shall see continued exploration of analogy as classically understood, culminating in the work of Richard Whately, and following him John Stuart Mill.

The First Introduction of the Concept of Analogy

Euclid V: Analogy and Incommensurable Magnitudes

The analogical relationship, interpreted as the identity of the relative magnitudes of two lengths $(A/B = C/D)$, first appears in Greek mathematics as an element in Pythagoras' theory of musical harmonics. However, for our purposes, we are concerned with a later use in which it becomes a key concept in Euclid V where it has a fundamental role in the solution to the problem posed by the discovery of incommensurable magnitudes, thereby marking a key stage in the development of mathematics.

The problem of incommensurable magnitudes arises as follows. We start out with the Pythagorean theory of relative magnitude, which says that we can specify the relative magnitude of any two lengths by two whole numbers (at this stage the phrase 'whole numbers' is pleonastic – the only numbers recognised are the positive whole numbers). Thus, for any two lengths, A and B, $(\exists m)(\exists n)(mA = nB)$, where, importantly, this formula can be given a straightforward geometrical interpretation: if you extend A to m times its length and B to n times its length, you arrive at two lines that are the same in length.

However, this simple theory received a death blow with the discovery of incommensurable magnitudes – the discovery that it was possible to construct a pair of lines for which it was impossible to satisfy this formula. The proof of this was a simple corollary of Pythagoras' theorem. Consider an isosceles right-angled triangle, with hypotenuse of length H, sides of length S. By Pythagoras' theorem, $H^2 = 2S^2$. If the Pythagorean theory of magnitude is correct, there exist two numbers p and q such that $H/S = p/q$ where p and q have no common factors, from which it follows that $p^2 = 2q^2$. Now, the square of an odd number is odd, and of an even number even, so that p must be even $= 2r$, say, giving us $4r^2 = 2q^2$, or $4r^2 = 2q^2$, giving us in turn that q must also be even, contradicting our assumption that p and q have no common factors.

This discovery constituted what may be regarded as the first crisis in the foundation of mathematics: it was now possible to specify lengths for which there could be no answer to the question of their relative magnitude. The task was thus to replace the Pythagorean theory of magnitude by one that was equally applicable to incommensurable and commensurable magnitudes. The two mathematicians who proposed solutions to this problem were Theaetetus,[2] and Eudoxus of Cnidus. It is the latter who concerns us here, and specifically the opening definitions in Euclid Book V that have been traditionally ascribed to Eudoxus. The most relevant definitions for our purposes are the following:

> Definition 3. Ratio is a mutual relation of two magnitudes of the same kind to one another in respect of quantity.
>
> Definition 5. The first of four magnitudes is said to have the same ratio to the second, that the third has to the fourth, when any equimultiples whatever of the first and third being taken, and any equimultiples whatever of the second and the fourth, if the multiple of the first be less than that of the second, the multiple of the third is also less than that of the fourth, and if the multiple of the first be equal to that of the second, the multiple of the third is also equal to that of the fourth, and if the multiple of the first be greater than that of the second, the multiple of the third is also greater than that of the fourth.
>
> Definition 6. Magnitudes which have the same ratio are called proportionals (analogous). When four magnitudes are proportionals (analogous),

[2] His solution to the problem is to be found in Euclid Book X.

it is usually expressed by saying the first is to the second as the third is to the fourth.

Definition 8. Analogy, or proportion, is the similitude (equality) of two ratios.

In the first of these, definition 3, where initially the notion of a ratio is left to be specified further in the subsequent definitions, the important thing to notice is the phrase 'of the same kind to one another' – you can talk of the ratio of one length to another length, or the ratio of one time to another time, but not of the ratio of a time to a length. It is the abandonment of this restriction when we come to extend analogy beyond its mathematical application that marks what is the most significant difference between the mathematical and the non-mathematical concepts of analogy.

The key definition is definition 5, which may be easier to understand if we render it in modern notation:

Given four magnitudes A, B, C and D, $A/B = C/D$ if and only if the following condition is satisfied:

$$(\forall m)(\forall n) \, ((m \times A > n \times B \to m \times C > n \times D)$$
$$\&(m \times A < n \times B \to m \times C < n \times D)$$
$$\&(m \times A = n \times B \to m \times C = n \times D))$$

where the quantifiers range over the natural numbers.

The strategy adopted here anticipates the strategy that was used in the nineteenth century to define real numbers. There you specify a real number by specifying which rational numbers are greater than it, which less, and which equal to it. Here you specify a relative magnitude by specifying which commensurable relative magnitudes are greater than it, which less, and which equal to it. The basic strategy here, which represents the high point of Greek mathematics and paves the way for modern mathematics, may be put as follows: it has been shown that we cannot in general specify the relative magnitude of any two lengths by citing a simple arithmetical formula for that magnitude, but we can nevertheless specify the relative magnitude of an arbitrary pair of lengths in the following sense. We can say when that relative magnitude is the same or different from the relative magnitude of any other pair of lengths. That is to say, we give the truth conditions of the formula $A/B = C/D$. This breakthrough was widely celebrated and led to a widespread interest in the concept of analogy, including, as we shall see, interest in the possibility of extending the concept beyond the realm of mathematics.

Of course, the central mathematical interest here is the way that it gives us a general theory of magnitude at an altogether more sophisticated level than the Pythagorean theory that it replaces. However, for our purposes, we are concerned with other features of the formula $A/B = C/D$ that emerge in the course of the subsequent mathematical investigations. Primarily we are interested in the way in which in Euclid Book VI this formula is used in the construction of analogical models and the development of a style of argument by analogy that will concern us throughout this book, including providing a clue to the form of the argument of the *Origin*. But before turning to that, there are two other features of the formula $A/B = C/D$, as defined by Eudoxus, which are also important for our purposes.

In the first place, one key characteristic of the idea of a ratio, as we have been looking at it so far, is the restriction contained in the phrase 'magnitudes *of the same kind*'. So that, for instance, it permits us to compare one length to another, one time to another, one volume to another and so on, but not one length to a time. However, we frequently in fact want to compare things in different categories. To take a simple example, to arrive at a concept of velocity requires us to compare the distance travelled to the time taken. It is the concept of analogy, as just explained in definition 5, that permits us to make sense of such comparisons between things in different categories. If we look again at that definition

$$(\forall m)(\forall n) ((m \times A > n \times B \rightarrow m \times C > n \times D)$$
$$\&(m \times A < n \times B \rightarrow m \times C < n \times D)$$
$$\&(m \times A = n \times B \rightarrow m \times C = n \times D))$$

we see that although we can only give a meaningful interpretation of it if A and B are 'of the same kind', there is no reason why A and C need be of the same kind. Thus whereas a theory of magnitude expressed purely in terms of ratios would make it impossible to compare a distance with a time, once we replace the Pythagorean theory with a theory expressed in terms of analogy, we can make sense of comparisons between things in different categories, and thus, e.g., find it possible to construct a concept of velocity, which precisely rests on comparing the distance travelled and the time taken. This leads us to an idea that, as we shall see, is stressed by Aristotle and indeed is crucial for making sense of subsequent applications of analogy, including those which we find in the *Origin*: *there are two different ways of comparing two entities A and B. The first, a direct comparison, only enables us to compare A and B if they are the same sort of thing, but the second, an indirect comparison resting upon an introduction of two other terms, C and D, and using the formula 'A is to C as B is to D', enables us to compare entities that, as Aristotle puts it, are remote, whether a distance to a time, an elephant's trunk to a hand, the opening chapters of a book to the opening shots in a battle, or a desert permitting only the most drought-resistant plants to survive to a racehorse owner permitting only the fastest horses to go to stud.*

The Alternation of Analogies

In the second place, we should note here one of the basic features of the analogical relationship that will turn out to have particular relevance to a full understanding of Darwin's argument for the competence of natural selection to explain the emergence of new species. *Analogies alternate.* That is to say, if A is to B as C is to D, then A is to C as B is to D. There is an elegant proof at Euclid V, Proposition 16, that this follows from the definition of analogy given in definition 5. The account of analogy throughout Book V has as its premise that the four terms of the analogy are 'of the same kind'. However, we shall later be concerned with an extension of this to cases where the terms of the analogy are not of the same kind. At this stage, we simply note that the possibility of alternating analogies with heterogeneous terms is exploited widely even within mathematics. Consider again the way in which we arrive at the concept of velocity. If body A travels a distance d_1 in time t_1, and body B d_2 in time t_2, where $d_1/d_2 = t_1/t_2$, then, alternating the analogy, the velocity of A, d_1/t_1, will equal the velocity of B, d_2/t_2.

The difference between the homogeneous case and the heterogeneous case is as follows. In the homogeneous case, we have the relation of A to B is the same as that of C to D, where all four terms are of the same kind. More explicitly ARB = CRD. What is proved in Euclid V is that in this case, ARB = CRD $\rightarrow$ ARC = BRD. However, in the heterogeneous case, because A and C are different in kind, ARC will typically make no sense. What we have is the weaker claim that there is a relation R' such that ARB = CRD $\rightarrow$ AR'C = BR'D, where it is determined contextually, on a case by case basis, what the appropriate value for R' is.

Euclid VI: Similar Triangles and Argument by Analogy

In Book V the basic properties of analogy, understood in its original mathematical sense of the equality of two ratios of lengths, were explored, in order to develop various applications of analogy in subsequent books.

For our purposes, what interests us is the way we can use the concept of analogy first to develop a concept of *an analogical model*, and then use that to explain a basic form of argument by analogy.

An analogical model may be explained as follows: suppose we have two domains of entities, and we set up a correspondence between entities in the one domain with entities in the other, thereby using the one domain as a model for the other. This model is an *analogical* model if there are a series of analogies between pairs of entities in the one domain and the corresponding pairs in the other. This is easiest to explain and understand by looking at the specific case that we find in Euclid Book VI.

Within Euclid, we are concerned with the simplest of all analogical models – the scale model, and, indeed the simplest of all scale models, two similar triangles. The central theme of Book VI is the theory of geometrically similar figures and of similar triangles in particular. Two triangles ABC and A′B′C′, with sides of lengths a, b, c and a′, b′, c′ will be similar if and only if $a/a' = b/b' = c/c'$: that is to say, if in accordance with the above definition, one is the analogical model of the other. We have made one triangle a model of the other by correlating the sides of the one triangle with the sides of the other. If the multiple instances of the analogical relation hold, then the triangles are similar, or, in other words, the first triangle is an analogical model of the second. Although this case is of extreme simplicity, its interest lies in the way that it can be readily extended to explain what it is for any two geometrical figures, of arbitrary complexity, to be similar to each other. The point of the concentration upon the case of similar triangles is that, since a triangle is the simplest rigid figure bounded by straight lines, a theory of similar triangles can readily be extended to a theory of geometrically similar figures in general: two figures will be geometrically similar if and only if every triangle inscribed in the first figure is similar to the corresponding triangle inscribed in the second. In this way we arrive at a general theory of what it is for one geometrical configuration to be a scale model of another.

We then proceed to prove that certain properties of the one triangle will be preserved by the modelling: that is, those properties will automatically be properties of the second, the most obvious such property being that corresponding angles of the two triangles will be equal, or that parallel lines correspond to parallel lines. Here we have a case of deductively valid analogical arguments, in which given that two geometrical figures are similar, we infer a range of additional properties that the two figures must have in common.

We can see the power of this style of argument by analogy if we reflect on one of its most familiar applications – the construction of maps of a terrain by triangulation. The map is constructed by creating a network of triangles on the page, each of which is similar to a corresponding triangle in a network of triangles on the ground. Once a map has been properly constructed in this way, the configurations on the map will share a wide range of additional topological features with the configurations on the ground. It is precisely this fact that gives maps their utility. Thus, for instance, when you say, 'These two dots on the map are separated by a blue line; therefore to get from this town to that town you must cross a river', you are in fact drawing a valid analogical inference (with, of course, the tacit premise that the map has been properly constructed).

What we have here is the basic form of a *valid* argument by analogy:

Domain A is an analogical model of domain B

F is a feature of domain A

Being F is invariant under analogy

∴ F is a feature of domain B.

What we need eventually to understand is what happens to such a pattern of argument when we transpose it from its mathematical setting to an empirical application. We can at this stage summarise the continuities and discontinuities involved in such transposition. We clearly have a valid argument form, regardless of its application. The difference lies in the question of the *soundness* of the argument, that is to say, the issue of the truth of its premises, and in particular whether we do have a genuine analogical model and whether the feature that interests us is indeed invariant under analogy. In the case of the similar triangles in Euclid, these premises are guaranteed a priori. We may simply posit at the outset that we are dealing with two similar triangles; we then go on to give a series of geometric proofs settling the question which features are indeed invariant under analogy. Once we move outside mathematics, arguments by analogy are only sound if we can give empirical support to, or other strong grounds for, accepting these premises. The reason that people are dubious as to the probative value of arguments by analogy is largely due to the widespread neglect on the part of those putting forward such arguments to give adequate grounds for believing precisely these premises. We shall seek to show that by contrast, Darwin's use of argument by analogy is fully responsible in this respect.

Archytas of Tarentum: Analogy and Definition

Almost all of the ideas necessary for the analysis of the argument of the *Origin* can already be derived from that which we have found in Euclid, and that argument is simply an application of the argument by analogy that we found in the case of similar triangles. However, within Euclid, the formula 'A is to B as C is to D' is always to be interpreted in purely mathematical terms – 'the ratio of A to B = the ratio of C to D'. For our purposes, we need to be able to give other, empirical, interpretations of this formula. We therefore need to look at subsequent developments in Greek thought to see whether and how we can extrapolate from the mathematical case to empirical interpretations of the analogical formula. Above all, we need to look at Aristotle's uses of analogy.

We look first at a use of analogy that is not prefigured in Euclid – the use of analogy to generate and define new concepts. For this, we need to consider Archytas of Tarentum. Archytas (428–347 BC) was a Pythagorean mathematician and statesman. Although only a few fragments of his writings survive, he is an important figure in the history of the concept of analogy. Not only was he a major influence on both Plato and Aristotle in their exploitations of analogy, it is in the slender evidence that survives that we find for the first time someone who is exploring the possibility of extending the concept of analogy beyond its original mathematical use.

His primary contribution, in the fragments for which we have evidence, is in the theory of definition, contained in these two quotations from Aristotle:

> Similarly, the consideration of similarities is useful for forming definitions that cover widely differing subjects, e.g., 'Calm at sea and windlessness in the air are the same thing' (for each is a state of rest), or 'A point on a line and a unit in number are the same thing' (for each is a starting point). Thus, if we specify the genus to be that which is common to all the cases, the definition may be regarded as appropriate. This is how those who frame definitions usually proceed: they state that the point is the starting point of the line, the unit the starting point of number. It is clear that they are assigning them both to the genus of what is common to the two cases.[3]

> It would seem that the definition by differentia is that of form and actuality, while that by constituent parts is, rather, that of matter. The same holds for the kind of definitions Archytas used to accept; for they are definitions combining matter and form. E.g., What is windlessness? Stillness in a large

[3] *Top*, I 108b23ff.

extent of air – the air is the matter, the stillness is the actuality and substance. What is a calm? Levelness of sea – the sea is the material substrate, the levelness, the actuality or form.[4]

Taking these two passages together, we may see Archytas as proposing a new way of defining concepts based upon analogy. We specify a concept not by noting properties belonging to all objects falling under that concept, but by grouping together as falling under a single concept a range of objects that are related analogically. Thus, to take one of Archytas' examples, the concept *calm*. We may talk of a calm sea, and a calm sky – and further of a calm mind, or of streets that are calm after a riot. These extremely heterogeneous entities have no obvious properties in common, but windlessness is to the sky as wavelessness is to the sea, as contentment is to the mind. Such concepts are extremely widespread, including *open*, *long*, *difficult*, *principle*, and typically permit us to group together objects that are different in kind (a long novel, a long pause, a long railway . . .). It is clear that Archytas' approach has far greater explanatory power for such concepts than the frequent superficial appeal to 'family resemblances'.

This represents a major advance in the theory of definition. The then standard account had been offered by Plato – the method of division. There you began with a class and then subdivided until you had specified the concept required, producing definitions such as 'man is a rational, sensitive, animate substance'. Although this is an excellent start in the theory of definition, it is extremely limited in its application and very few scientifically fruitful concepts can be defined by this means. By contrast, a wide range of important concepts, both within and outside science, are susceptible of being explained along the lines outlined by Archytas. A major part of the greater power of such definitions – the part that will be stressed by Aristotle – is the capacity of such definitions to gather together and make scientifically significant comparisons between highly disparate phenomena.

When we come to the *Origin*, we shall find that two of Darwin's central concepts, 'struggle' and 'select', are paradigm cases of concepts that are best handled by Archytas' method.

Plato: The Informal Use of Analogy

Plato was a friend of Archytas, and was clearly familiar with his work.[5] There is, however, not the rigorous use of analogy that we find either in

---

[4] *Meta*, VIII 1043a 22ff.
[5] For instance, the reference in "'We may venture to suppose", I said, "that as the eyes are framed for astronomy so the ears are framed for the movements of harmony; and these are in some sort kindred

the mathematics or that we shall encounter in Aristotle. There is also no theoretical discussion of the concept. What we find instead are several informal arguments by analogy based on an intuitive use of analogical models, such as 'the cave'[6] and 'the line'.[7] By looking at one, the most famous of these, we can see how we may transpose the idea of argument by analogy to a non-mathematical setting. This argument may be used to give us a preliminary indication of the form such an argument should take. Also, as it stands, Plato's argument is at best suggestive, and by seeing why it falls short of being a fully rigorous argument, we may use it to identify those features that would be required of such arguments by analogy to make them watertight.

The argument that we shall consider is the extended argument that constitutes the *Republic*. This argument is intended to demonstrate that it is better to be just than unjust. It is put forward as a reply to a challenge made by Glaucon and Adeimantus, who argue that, the nature of the world being such as it is, it is the perfectly unjust man who flourishes, but not only that, since he needs to appear to be just to carry out his nefarious schemes, he will become a benefactor of humankind. By contrast the perfectly just man will be condemned to a life of misery, and since he will not be concerned to *appear* to be just, will be constrained from behaving in ways that are to the obvious benefit of the rest of humankind.

The argument rests on using the state as an analogical model for the soul. We are then invited to infer that the ideal state having such-and-such a structure, so too the ideal for the soul will have an analogically corresponding state. The state is seen as having three components – a ruler, a group whose task is to ensure that the people obey the laws of the ruler, and the people. In the ideal state, the ruler will be the philosopher king, who knows 'the form of the good' – what is right should be done, and then there are guardians who have true beliefs about what should be done and whose task is to make the general public carry out the wishes of the king. It is then argued that the soul has also three components – reason, a 'spirited' part, and appetite. This is shown by the fact that we are sometimes torn between different courses of action, and in particular between what reason dictates that we ought to do, and what our appetites tell us that we want to do, and, further that in such cases we can by an act of will control our appetites. This enables us to use the state as an analogical model for the

sciences, as the Pythagoreans affirm and we admit, do we not, Glaucon?"' (*Republic*, 530d) has been ascribed to Archytas.
[6] *Republic*, 514a–520a.   [7] *Republic*, 509d–511e.

human soul. We then argue that just as the best state – the just state – is one in which the philosopher king is in power and the guardians ensure his laws are obeyed, so the best soul – the just soul – is one in which reason always controls the passions.

There is clearly much that is questionable about this argument. Is a society in which the majority of the people have no control over their lives and are simply made to do what the ruler dictates really an ideal society? However, we are concerned with the question how well the argument stands up as an analogical argument. That will give us a preliminary indication of what needs attending to in assessing *any* purported analogical argument. Even if we grant Plato his premise that he has shown the republic to be the just society – that is not our present concern, which is whether if we accept that as a premise, he is entitled to infer his conclusions as to what it is for the soul to be just. If we were to accept Plato's argument we would clearly need to be satisfied about two things. Firstly, we would need to ask whether we have here a genuine case of an analogical model. That is to say, is it really the case that the relation between a ruler and the citizens was the same as the relation between reason and appetites in the soul: that the ruler: the citizens :: reason: appetite? To be convinced of that we would need to accept Plato's argument that the best way to explain being torn in different directions as to what to do is to hypostasise three different aspects of the soul, in such a way that we can take seriously the idea of one of these aspects governing another. Secondly, once we had established that there was indeed an analogical model here, we would then need to show that being just was a feature that was invariant under analogy – that if the state were just, it would follow that a soul with the corresponding structure would also be just. It is here perhaps that the argument is most vulnerable: there is something like an equivocation on the word 'just' here. We are talking about something very different when we describe a state as just and when we talk about an individual as just. For a state to be just is for there to be just treatment of the citizens within the society, for instance, for there to be equality before the law; however, for an individual to be just concerns their relations with other people: in one case we are concerned with the internal relations within the state, in the other with the external relations of the individual.

Aristotle: Two Ways of Comparing Things

It is in the use that Aristotle makes of the concept of analogy, rather than the more informal uses that we find in Plato, that we find the full potential of the concept once it is extended beyond its mathematical origins. Aristotle shows how appeals to analogy can be made to do important work in research in the varied contexts – in rhetoric, in his researches into living things, in metaphysics, in the theory of justice and in the theory of definition. In each of these cases, he is typically precise and rigorous, with a surefooted understanding of what analogy can, and cannot, do for us. However, for our purposes, we shall not look in detail at each of these particular applications: only two of these applications will prove to be of importance in eventually understanding the role of analogy in the *Origin*: that of the relation of analogy to metaphor, and that of the question of method in biology.[8] We shall look at Aristotle on metaphor in the course of Chapter 6, and before looking at the biological writings, we shall identify and examine some of the themes that recur in all these applications.

We begin with what may be regarded as a key theoretical statement: 'Yet a further method of selection is by analogy: for we cannot find a single word applicable to a squid's pounce, fish spine and bone, although these too possess common properties as if there were a single osseous nature.'[9] In a way this is simply a repetition of what we found in Archytas, but it is worth interpreting it in the context in which Aristotle introduces the idea. The major part of the *Posterior Analytics* is concerned with definition by the method of division, where a definition proceeds by taking a large class, dividing it, then subdividing until you have specified the class of things that interest you, producing such definitions as 'Man is a rational, sensitive, animate substance'. He is working here against a Platonic background, in which there were ways of 'carving nature at the joints',[10] producing natural kinds, and where natural kinds could always be defined by the method of division.

But at the same time, even at the stage of his enquiry represented by the *Posterior Analytics*, there is something that will assume central significance in the *Parts of Animals*: even within the science of biology there is a need for concepts that do not pick out natural kinds. He proposes that such concepts can be defined by analogy, along the following lines. We start by defining the species of animals by the method of division, setting up a Porphyry tree branching downwards, with the different species at the bottom. We then cut across the tree picking out functionally

---

[8] The word 'biology' is of course a much later invention, but as with most Aristotelian specialists, we may apply it to Aristotle as giving an accurate picture of the nature of his investigations.

[9] *PostA*, II. 98a 20   [10] *Phaedrus*, 265e.

corresponding parts of different animals. In the example he gives, we take the example 'bone', where we have monkey bone, fishbone and cuttlebone. These are composed of completely different substances and thus do not form a natural kind. They are however analogically related: monkey bone is to monkeys as fishbone is to fishes and as cuttlebone is to cuttlefish, in that in each case we are dealing with substances that can provide the skeletal structure for their host. In order to provide such a skeletal structure, they must share those properties, such as hardness, that are necessary or useful in performing their function. In this way, we arrive at an 'as-if' natural kind, which can be the subject of scientific investigation.[11]

> Likenesses must be studied between things in different genera, the formulae for such likenesses being 'As *A* is to *B*, so is *C to D*', such as 'As knowledge is to what is known, so is sensation to what is sensed', and also 'As *A* is in *B*, so is *C* in *D*', such as 'As sight is in the eye, so is reason in the soul', or, 'As wavelessness is in the sea, so is windlessness in the air'. In particular, we must have practice in comparing genera which are remote; for in the other cases, the similarities will be more readily apparent.[12]

The governing idea that runs through all the diverse applications that Aristotle makes of the concept of analogy is that there are two fundamentally different ways of comparing two things, A and B: there is, firstly, making a direct comparison, which is a matter of noting common properties of A and B, and secondly, making an indirect comparison, in which we introduce third and fourth terms, C and D, such that A is to C as B is to D. This contrast is highly flexible and may be adapted in a way that is appropriate to the topic of research. Thus, Aristotle shows how exploitation of this contrast can illuminate questions concerning the nature of justice, explain the difference between successful metaphors and lifeless metaphors, be fundamental to comparative anatomy and throw light on the question what could be meant by equality in the state. In each of these examples, the analogical relation 'A: B :: C: D' is to be understood in a way that fits the topic in hand. In every case apart from metaphor, it is clear how the formula is to be interpreted and Aristotle holds rigorously to that

---

[11] Given that we now, unlike Aristotle, believe that biological entities are the products of a partly random and chaotic process, it becomes even more urgent to explain the possibility of biology as a science when few, if any, of its concepts, such as 'species' or 'law' pick out natural kinds. It is at least worth exploring the question whether Aristotle's proposal that analogy can be used to explain such 'as-if' natural kinds can be worked out.

[12] *Top*, I. 108a 6ff.

interpretation.[13] Thus in the case of justice and equality in the state, Aristotle contrasts arithmetical equality – everyone is paid the same – and analogical equality in which everyone is paid proportionally to their role within society; in the case of metaphor, we have a contrast between metaphors based upon a transfer from species to species or genus to species and metaphors based on analogy; and in his researches into living things there is a contrast between parts of animals that differ by the more and the less and parts of animals that are only related by analogy.

One element in the quotation we are now looking at is the idea that analogical comparison, unlike a direct comparison, permits us to compare 'genera that are remote'. Here the word 'remote' should be read as broadly as one chooses: no matter how different in kind two entities may be, an analogical comparison between them may still be possible. We already saw this when we looked at Euclid, showing how Eudoxus' definition of analogy permitted us to arrive at a concept of velocity by comparing time taken with distance travelled. When we move outside the mathematical context, everyday examples clearly show this to be true. An opening batsman at cricket, the opening chapter of a book, a chess opening and the opening of a new hospital: here we have four entities that share no obvious intrinsic properties that can nevertheless be fruitfully seen as all openings because of the way that they are related to what is to follow. Aristotle himself will show how this can be exploited, e.g., in his theory of justice.[14] Suppose you wish to explain what a fair price is for something. You constantly have to compare things that are remote – the price of a house and the price of legal representation in court. A direct comparison of goods that are so different is virtually impossible. Aristotle therefore proposes that we explain a fair exchange by finding an interpretation of the analogy: this much legal representation is to the lawyer as a house is to the builder – e.g., how much it costs each to produce their respective goods.

Finally, Aristotle is claiming that it is the analogical comparisons, as opposed to the direct comparisons, that will characteristically be the scientifically fruitful comparisons. Thus when we come to look at Darwin's use of analogy, whereas comparing the activities of one breeder with another simply tells much the same story, it is the comparison between a farmer castrating a bull and frost killing a lettuce that leads to the idea of natural selection.

---

[13] In the case of metaphor, we are dealing with an essentially unsystematic and opportunistic use of analogy.

[14] See, e.g., NE, 1133a 6, 11

Analogy in Biology

Aristotle's contribution to the early history of biology is one of his most assured accomplishments with a deep influence on all his successors until Darwin, so that even someone as late as Cuvier will explicitly describe himself as applying Aristotle's methods in his work. Darwin himself writes in such a way as to indicate that he regarded Aristotle as the greatest of his predecessors.[15] There is a widespread opinion that Darwin's work made Aristotle's work obsolete. However, that is a simplistic reaction. What is clear is that Aristotle had argued for the fixity of species, and as a result could only account for the high level of functional complexity evident in animals, by positing as an underlying metaphysical principle that nature itself was purposive, leading to the methodological principle that 'nature does nothing without purpose or makes anything superfluously'.[16] The theory of evolution by natural selection enables us to replace the highly counterintuitive metaphysical principle that Aristotle found himself forced into by a simple, purely naturalistic interpretation of the purposiveness that is evident throughout the animal kingdom.

However, even if the metaphysical background to Aristotle's account of animals and plants may be regarded as refuted by modern evolutionary theory, that leaves the method for biological research that he argues for untouched. As a pioneer, working with limited empirical information, it is inevitable that Aristotle will frequently go wrong in the conclusions that he draws. However in subsequent centuries, further empirical investigations making more refined application of Aristotelian methods are responsible for a large proportion of the advances in pre-Darwinian biology, such as, for example Harvey's discovery of the circulation of the blood.

Unlike previous explorations in biology, and indeed his own investigations in the *Posterior Analytics*, in his biological writings properly speaking, Aristotle shows little interest in the question of definition. At *Parts of Animals* 642b 5ff., he makes a long series of devastating criticisms of the attempt to define animals by the method of dichotomous division, and that is all. Instead, he sets himself a different task, that of discovering why animals are the way they are, where this question must be interpreted in terms of discovering their 'final causes'. That is to say, the kind of

[15] Letter to William Ogle, February, 1882: 'From quotations which I had seen, I had a high notion of Aristotle's merits, but I had not the most remote notion what a wonderful man he was. Linnaeus and Cuvier have been my two gods, though in very different ways, but they were mere schoolboys to old Aristotle.' For the relation of Darwin to Aristotle, cf. Gotthelf (1999), pp. 3–30.

[16] *PA*, II 651b 24.

explanation he is after is illustrated by the reply to the question 'Why do giraffes have long necks?', 'So that they can browse the leaves of tall trees'.

To understand how he is to set about offering answers to such questions, we must first look at the way that he is thinking of animals. They are for him the paradigm case of *substances*, entities that are systems of parts, whose continued existence depends upon the cooperative activity of those parts. They are composed of matter, the stuff they are made of, and form, the principle of organisation of the parts to produce a functioning whole. In the case of living substances, the form of the being is what Aristotle calls its 'soul'.

This conception is clarified by the simple observation that Aristotle makes at the beginning of the *Parts of Animals*. There is a range of things that an animal needs to do if it is successfully to live out its life (and reproduce): 'For genera that are quite distinct still frequently present many identical phenomena, sleep, for instance, respiration, growth, decay, death, and other similar affections and conditions, which may be passed over for the present, as we are not yet ready to treat of them with clearness and precision'.[17]

If an animal can perform such actions, it must be assembled in such a way that enables it so to do. A part may then be identified functionally, as *organ*,[18] as the feature of the animal that gives it the ability to do certain things.

In the light of this, Aristotle's question now becomes the questions 'Why does an animal have the parts it does?' and 'Why do these parts have the form they have, and why are they related the way they are?' To answer such questions, Aristotle makes what is perhaps his most significant contribution to the theory of life, the use of comparative biology. One compares corresponding organs in different species of animals. The similarities between these organs give one a guide as to what is necessary for such an organ to carry out its function, or at least carry out that function well. The differences give one a guide as to the way that such organs are adapted to fit the life of the particular animal.

This enquiry is governed throughout by a major contrast between the ways in which corresponding organs are related: they either 'differ by the more and the less' or are 'related only by analogy'. The term 'differing by the more and the less' stems from Plato[19] to signify cases where two things possess the same property but to different degrees. Two things will thus

---

[17] *PA* 639a, 19–23   [18] 'Organon' is Aristotle's metaphor. The word literally means a tool.
[19] *Philebus*, 24a–25d.

'differ by the more and the less' if all the differences between them can be specified simply by the use of comparatives: one is longer, heavier, hotter . . . than the other. We can see the straightforward biological application of this idea if we consider the case of beaks. Given any two birds, when we compare their beaks, we find basically the same structure, composed of the same stuff, but in the one case the beak is longer, straighter, more pointed . . . than the other. By way of contrast, if we consider the case of the horns and tusks of different animals, we have the same kind of organ of defence, but with far greater differences than can be specified simply by the use of the more and the less. These organs are related by analogy, in that they have the same function: horns are to bulls as tusks are to elephants.

The two different ways in which organs are related play opposite roles in Aristotle's investigations. In the case of parts that differ by the more and the less it is the *difference* between the parts that are significant but in the case of parts that are related only by analogy, it is the *similarities*.

In the case of parts that differ by the more and the less, the differences show one the way in which the parts have been fine-tuned to fit the way of life of the various animals:

> Various sorts of beak are found, to suit the various uses including defence to which it is put. All of the birds known as crook-taloned have a curved beak, because they feed on flesh and take no vegetable food: a beak of this form is useful to them in overcoming their prey, as better fitted for the exertion of force. . . . Every bird has a beak which is serviceable for its particular mode of life. The woodpeckers, for instance, have a hard beak . . . small birds, on the other hand, have finely constructed beaks, for picking up seeds and catching minute animals.[20]

In the case of parts related by analogy, despite the fact that they are related only by analogy, they will typically also possess a range of common intrinsic properties. These are the properties that an organ of that kind must have if it is to function, or at least function well. Thus if we look at the case of an elephant's tusk and a bull's horn, they are, for instance both curved, pointed, of the same texture and attached to the head. By reflecting on such common properties as our investigations identify, we can further our understanding of the part and its function.

> Again, Nature acted rightly in placing the horns on the head. Momus in Aesop's fable is quite wrong when he finds fault with the bull for having his horns on the head, which is the weakest part of all, instead of on the shoulders,

---

[20] *PA* III 662b, 1–10.

which, he says, would have enabled them to deliver the strongest possible blow. Such a criticism shows Momus' lack of perspicacity. If the horns had been placed on the shoulders, as indeed on any other part of the body, they would have been a dead weight, and would have been no assistance but rather a hindrance to many of the animal's activities. And besides, strength of stroke is not the only point to be considered: width of range is equally important. Where could the horns have been placed to achieve this? It would have been impossible to have them on the feet, knees with horns on them would have been unable to bend; and the bull has no hands; so they had to be where they are – on the head. And being there, they offer the least possible hindrance to the movements of the body in general.[21]

In implementing his programme, Aristotle uses these two ways in which corresponding parts of animals are related to divide the animal kingdom into nine major families along the following lines:

Some may find it surprising that everyday usage has not combined the water-animals and the feathered animals into a single group, and adopted one name to cover both, since these two groups have certain features in common. The answer is that in spite of this the present classification is the right one, because while groups that differ only 'by excess' (that is, 'by the more and less') are placed together in a single group, those which differ so much that their characteristics are only analogous are separated out groups. For example: one bird differs from another bird 'by the more and less', or 'by excess': one bird's feathers are long, another's are short; whereas the difference between a bird and a fish is greater, and their correspondence is only by analogy: a fish has no feathers at all, but scales, which correspond to them.[22]

Then, in the *History of Animals*, Aristotle conducts an extensive and meticulous survey of the parts of animals working systematically through the different parts that he has identified, noting in each case the similarities and differences between the corresponding parts in different species of animals. Against that background, he will then in Books II to IV of the *Parts of Animals* attempt explanations of what his survey has shown.

We may conclude this survey by seeing how such explanations look in the particular case by taking a case where both kinds of comparison are combined in a single part. On the one hand, the elephant's trunk differs by the more and less from the human nose, and on the other is related by analogy to the human hand:

For the most part, there is very little variation in the organ of smell among the viviparous quadrupeds. ... In the elephant this part is unique in its

[21] *PA* III 663a 34–663b 12.   [22] *PA* I 644a, 12–23.

extraordinary size and nature. Using its nostril as though it were a hand, the elephant conveys food, both solid and liquid, to its mouth, and it uses it to uproot trees by winding it round them. In effect it puts it to all the uses to which a hand is put. The reason for this is that the elephant has a double character, both as a land-animal and as a water-animal. It needs to get its food from the water, but at the same time must breathe, being a blooded land-animal. However, because of its great size, it cannot move rapidly from water to land, as do some other blooded vivipara that breathe. Thus it needs to be equally at home on land and in the water. In the same way, then, that divers are sometimes equipped with an instrument for breathing, giving them access to air from the surface while they are under water, so that they may remain for a long period under the sea, nature has provided the elephant with its elongated nose.[23] Whenever they cross deep water, they lift their trunks up to the surface and breathe through it. For, as I have already said, the elephant's trunk is actually a nose. Now it would not have been possible for the nostril to discharge all these functions if it had not been soft and pliable. For then its sheer length would have prevented it from feeding, in the same way that the horns of certain oxen do, so that they are obliged to walk backwards while grazing. Thus it is soft and flexible, and because it is such, nature has, in her usual way, exploited this by assigning to it an extra function as well as its primary one – it performs the function of forefeet. In polydactylous quadrupeds, the forefeet do not merely support the animal; they serve as hands. But elephants (which, having neither a cloven hoof nor a solid hoof, must count as belonging to this group) are so huge and heavy that their forefeet are reduced to mere supports; and, indeed, because they move so slowly and bend with such difficulty, they are quite unfit for any other purpose. A nose, then, is given to the elephant for breathing, in the same way that one is given to every lunged animal; this is at the same time elongated and capable of being coiled round things, because the elephant spends so much of the time in the water, and takes time to move onto dry land. And since the forefeet are unable to fulfil the normal function of forefeet, nature, as I said, assigns to this part the rôle of discharging the function that should have been performed by the forefeet.[24]

Here we see the peculiarities of the elephant's trunk explained in terms both of its interrelationship with the other parts of the elephant and of its life and environment. Because it needs to be able to wade across water, it needs to breathe when it is submerged. Because of its bulk, it cannot readily bob in and out of the water in the way that, say, otters do. Therefore its breathing organ is an elongated flexible tube which can be used as a snorkel. At the same time, because of its bulk its legs must be like

[23] For modern evidence that Aristotle was right on this point, see West (2001), pp. 1–8.
[24] *PA* II 658b 2–659a 37.

pillars. Therefore it is unable to provide itself with forepaws. Therefore it exploits its trunk, as a long and flexible tube, shaped in such a way that it can form an analogy of an arm and hand.

What this example shows, as do many of Aristotle's examples, both in *The Parts of Animals* and elsewhere, is the way in which Aristotle's quest for teleological explanation leads to what can be described in modern terms as a sophisticated form of 'adaptationism' – an interplay between three different forms of adaptation: the way the parts of animals are adapted to the animal's environment, the way they are adapted to its way of life, and the way the parts of the animal are adapted to one another. Of course, Aristotle's understanding of such adaptation was quite different from post-Darwinian adaptation developed in the context of a theory of natural selection: for Aristotle this was, rather, a consequence of the 'metaphysical' principle that 'Nature always works for the best'. What is also dubious is his apparently taking the further step of thinking that there must be a similar teleological explanation of why a particular species of animal exists.

Although the method of comparative biology that he has evolved depends upon comparing animals that are related only by analogy, the resulting arguments are not arguments by analogy in the sense that interests us in this book, but a form of what we would now call 'arguments to the best explanation'. Thus, Aristotle asks 'Why, in every species, no matter how morphologically different their eyes, are the eyes always located in the head of the animal?', and goes on to argue that this is explained by the fact that it is this placing of the eyes that is best suited for them to carry out their function in enabling the animals to take in their surroundings.

Argument by Analogy

Despite the extensive explorations of the concept of analogy that we find in Aristotle, *arguments* by analogy are comparatively uncommon. The clearest cases that we find are those in which he exploits analogies between animals and other complex entities such as the city state or, even, in the *Poetics* tragedy. In particular, he is concerned with the question 'In what way are such entities integrated and united as a single thing?'

If we return to Aristotle's conception of animals as substances, on his account animals are united as a single entity in a way that is quite different from the way in which a rock, say, is a single thing. In both cases we can talk of them as entities with parts, but for Aristotle this is so in completely different ways. A rock is a continuous lump of uniform material with a surface separating what is inside from what is outside the rock and a *part* of the rock is any continuous lump of material wholly contained within the surface of the rock. However, when Aristotle talks of the *parts* of substances, such as animals, he conceives of such parts quite differently. Here the parts are identified by their *function* in enabling the whole to flourish: eyes are the part that enables a man to see, etc. An animal is then to be seen as a system of such parts, organised in such a way as to permit the animal successfully to live out its life cycle. The way the parts are organised for this to be possible is the form of a substance. In the case of a living being this is its 'soul' (ψυχή): that form of organisation that permits it to live.

In this way, an Aristotelian substance is dependent upon the proper functioning of its parts for its successfully continuing to thrive. The contrasts between such a substance and another entity such as a rock and between an Aristotelian part and a material part is sharp. What unifies a rock as a single entity is its being a continuous piece of matter, with clear boundaries. What unifies an animal as a single entity is its being a system of parts that cooperate to produce something that can live. If a material part is cut out of a rock, it simply continues to exist as a smaller rock. If an Aristotelian part of an animal is removed or damaged, the whole animal is impaired or disabled. Anything that can be removed or damaged without affecting the whole animal is not to be counted as a part of that animal.

In the light of this account, we can see how Aristotle models his account of the structure of tragedy, and in particular what it is for a tragedy to be a unified whole by transferring elements of that account. Having first argued for the centrality of plot in his account of tragedy, at 1450a 39 he establishes an analogy between a tragedy and a living organism as follows: 'So the plot is the source and (as it were) the soul (οἷον ψυχή) of tragedy.'

Aristotle will then pursue this analogy in his accounts of both the 'magnitude' and above all the unity of a tragedy:

> Any beautiful object, whether a living organism or any other entity composed of parts, must not only possess those parts in proper order, but its *magnitude* also should not be arbitrary; beauty consists in magnitude as well as order. . . .[25]

> . . .

> A plot is not (as some think) unified because it is concerned with a single person. An indeterminately large number of things happen to any one person, not all of which constitute a unity. . . . Just as in other imitative

[25] *Poe* 1450b, 35–38. Translation from Malcolm Heath, *Aristotle, Poetics*, Penguin Books Ltd., London 1996.

arts the imitation is unified if it imitates a single object, so too the plot as
the imitation of an action, should imitate a single, unified action – and one
that is also a whole. So the structure of the various sections of the events
must be such that the transposition or removal of any one section dislocates
and changes the whole. If the presence or absence of something has no
discernible effect, it is not a part of the whole.[26]

We see here Aristotle directly redeploying by analogy principles from his
account of living organisms in his analysis of tragedy: that which makes a
successful tragedy being understood by analogy with that which makes a
healthy animal.

There is a major problem still to address. Let us suppose that we have
established that A is an analogical model of B, and that A has feature F. How
do we know whether F is one of the features of A that can be transferred to
B? Within Euclid answering this question was straightforward. A series of a
priori geometrical proofs can prove which features of similar triangles must
be shared, and which need not be. However, once we consider arguments by
analogy, no such a priori proofs are possible, and need replacing with
empirical answers to this question. What makes one uneasy about Plato's
use of analogy is that he nowhere addresses this issue.

In the case of Aristotle, there is one passage which needs to be considered
in this connection, with two reservations: although the argument he considers can clearly be presented as an argument by analogy, he himself
describes it as 'παράδειγμα' ('example')[27] and also it is a passage whose
interpretation is controversial. This is *Prior Analytics*, II/24 (68b 39–69a 19).

> Let A be 'bad', B 'to make war on neighbours', C 'Athens against Thebes'
> and D 'Thebes against Phocis'. Then if we require to prove that war against
> Thebes is bad, we must be satisfied that war against neighbours is bad.
> Evidence of this can be drawn from similar examples, e.g., that war by
> Thebes against Phocis is bad. Then since war against neighbours is bad, and
> war against Thebes is against neighbours, it is evident that war against
> Thebes is bad. Now it is evident that B applies to C and D (for they are
> both examples of making war on neighbours), and A to D (since the war
> against Phocis did Thebes no good); but that A applies to B will be proved
> by means of D.

This is extraordinarily compressed. Cast as an argument by analogy it
would run: 'War by Thebes against Phocis is bad. Thebes is to Phocis as

---

[26] *Poe* 1451a, 16–35.

[27] Aristotle presents 'example' as a type of proof that is neither deductive (syllogistic) nor inductive,
since it is an inference from a single particular situation to another different particular situation.

Athens is to Thebes (both being neighbours). Therefore war by Athens against Thebes is bad.' If this argument is to go through, we need to consider Aristotle's justification: 'we must be satisfied that war against neighbours is bad. Evidence of this can be drawn from similar examples, e.g., that war by Thebes against Phocis is bad.' As it stands, this is inadequate, and fails to do justice to what Aristotle has in mind in his discussions of 'example': we are meant to discern a general truth in a clearly understood particular case. However, the facts that Thebes waged war against Phocis, that Thebes and Phocis were neighbours and that this turned out badly for Thebes does not as yet suggest the general truth that going to war against a neighbour will turn out badly. To arrive at the general truth, we need also what Aristotle fails to capture, some link between the fact that Thebes and Phocis were neighbours and that the war turned out badly: what we need is for the example to show *why* it is bad to make war against a neighbour. The task of replacing Aristotle's inadequate formulation with something more precise lies ahead of us. In fact it is not until the end of the eighteenth and beginning of the nineteenth centuries that we find attempts being made to resolve that task, and we shall be turning to that towards the end of the next chapter.

Retrospect: Analogical Models and Argument by Analogy

Analogy within Mathematics

We may conclude this chapter by looking again at the most important feature of the use of analogy for our purposes: the beginnings of the idea of argument by analogy as understood in the context both of classical Greece and, we shall argue, the interpretation of Darwin's argument in the first four chapters of the *Origin*. Many facets of the concept of analogy that we have encountered in this chapter will recur throughout this book – such as, for example, Aristotle's stress on the capacity of analogical comparisons to compare 'things that are remote' (a drought is related to the plants in its region as a farmer is to the livestock in his care). However, our central concern is specifically with the concept of *argument* by analogy and the related idea of an analogical model. We shall therefore trace through what happens to *these* concepts in the period we have been looking at.

We may, somewhat artificially, divide the development of the examination of the concept of analogy during this period into three stages: the first stage is the analysis of the concept itself; the second stage is the extension of the concept within mathematics to give us the idea of an analogical model and the rigorous treatment of the simplest form of argument by analogy; and the third stage is the exploration of the question how far the ideas that have been developed within mathematics can be transferred into the empirical domain.

For our purposes, we may regard the first stage as Eudoxus' solution to the problem of incommensurable magnitudes. The discovery that there were cases where the relative magnitude of two lines could not be expressed as the ratio of two whole numbers meant that a different approach to the question of magnitude was necessary. Eudoxus' solution was to replace the attempt to express the relative magnitude of two lines by means of a closed arithmetical formula, by instead giving an account of the conditions under which the relative magnitude of two lines A and B equals the relative magnitude of two lines B and C. That is to give the truth conditions of the formula $A/B = C/D$. This is given a precise account in Euclid V, definition 5, and in definition 6 such a relation is called 'analogous'. This understanding of the analogical formula is fixed at the outset, and will remain constant throughout the enquiry. Book V will then explore further properties of the analogical relationship, such as proving that analogies alternate (If $A/B = C/D$ then $A/B = C/D$.) This clearly provides a *general* theory of magnitude, applicable not only to lines, but also to areas, volumes, times, etc.

The next stage builds in Book VI on the account of analogy in Book V, to give us the most straightforward case of argument by analogy in the theory of similar triangles. We have here two triangles ABC and $A'B'C'$ such that there is a series of analogical relations between the sides of the triangles: $AB/A'B' = AC/A'C$, $AB/A'B' = BC/B'C'$ and $AC/A'C' = BC/B'C'$. This gives us the simplest of all cases of analogical objects: two domains of elements, with a series of analogies between the various elements in one domain and the corresponding elements in the second domain. We can now construct a series of arguments by analogy, proving which properties are invariant under analogy – which may be transferred from the model to its target, and, equally importantly, which may not. (If two lines *l* and *m* in the model are parallel, then the corresponding lines *l'* and *m'* in the target will be parallel; corresponding angles will be equal; however, the size of areas, say, will not be preserved.) What we have here is a paradigm case of a fully valid set of arguments by analogy, where it is possible to develop purely a priori arguments to show what is and what is not invariant under analogy. This account can easily be extended from similar triangles to any geometrical configurations whatever, giving us a general theory of scale models.

Analogy beyond Mathematics

In the final stage, possibly beginning with the Pythagorean mathematician, Archytas of Tarentum, there are explorations of the possibilities opened up if we extrapolate the idea of analogy from the mathematical to the empirical domain. For this, we take the formula 'A is to B as C is to D' – or more explicitly, '(ARB = CRD)' but now interpret the formula, according to context, with a variety of different accounts of the relation 'R'. This possibility was taken up and developed, even if in very different ways by both Plato and Aristotle.

Plato

We find extensive use of analogies in Plato. For instance, in the *Republic*, we have 'the divided line'[28], 'the cave'[29], and the overarching analogy of the *Republic* itself as an analogical model for the tripartite soul. Many of these are difficult to interpret in detail, and many of them are put forward purely to illustrate an idea rather than present arguments by analogy. The one that is clearly presented as an *argument* is the comparison between the ideal republic and the tripartite soul, aiming to show that it is better for someone to be just than unjust. However, these arguments are at best suggestive, and it always remains controversial whether Plato is justified in thinking that there is a genuine analogy that would justify the inference that we are invited to make.

Aristotle

We find a far more disciplined use of analogy in Aristotle. Three ideas dominate his discussions, each of which can be seen as having some relevance for our discussions throughout this book. There is a stress on the difference between analogy and simple similarity, with the claim that analogical comparisons are 'scientifically' more illuminating than comparisons based on common intrinsic properties of two things, since the latter comparisons are typically trivially obvious. Next there is an emphasis on the idea that it is analogy that gives us the capacity to make significant comparisons between 'things that are remote' (e.g. in different categories). Finally, Aristotle shows by example the way in which analogy can be used to throw light on or solve problems in a wide variety of fields, ranging from

---

[28] *Republic*, 509d–511e.   [29] *Republic*, 514a–520a.

justice to metaphor to biology – in general, wherever we want to explore the relation between 'incommensurable' entities.

Argument by analogy is far less frequent than in Plato – the most obvious examples being his application by analogy of his account of substance to other complex entities such as the state or tragedy, and scattered remarks – mostly in the *Organon* – concerning what he calls 'παράδειγμα' ('example'). At *Prior Analytics*, II/24 (68b 39–69a 19), he gives what may be regarded as the first attempt to give a theoretical account of the form of such arguments.

CHAPTER 2

Analogy in the Background to the Origin

Beginning with the Aristotelian revival in Islam, which was then followed in Christian scholasticism, the mediaeval period is one of the richest and most interesting periods for anyone seeking to understand analogy and its uses. What we find are discussions that are typically more sophisticated and theoretically self-conscious than the overwhelming majority of discussions nowadays. In a general history of analogy leading up to Darwin, omission of this period would be incomprehensible. However, for the purposes of the present book looking at the mediaeval discussions of analogy would be a red herring, distracting attention from what, in our context, is important, for two main reasons. In the first place, the major concerns throughout the mediaeval period are, on the one hand, onto-logical, the nature of the analogy between God and the world ('the *analogia entis*'), and, on the other hand, linguistic, appeals to analogy to explain the possibility of using human language to talk about God without falling into anthropomorphism. What we do not find in this period is any concern with what is important to us in this book: *argument* by analogy,[1] neither looking at examples of such, nor exploring the idea theoretically. In the second place, there is a shift in terminology in this period: mediaeval *analogia* does not correspond exactly to Aristotle's 'ἀναλογια'. Aristotle had talked of 'words said in many ways', by which he meant cases where it was appropriate to say both that 'A is F' and 'B is F', but not in virtue of A and B sharing an (intrinsic) property. These included cases that Aristotle would have counted as 'said by analogy' (the opening chapter of a book and the opening moves in a chess game). In particular, he also included what eventually came to be called 'the analogy of attribution';[2] these were

---

[1] In particular, we know of no instance of someone in this period putting forward what became prominent in much post-Enlightenment apologetics, an argument by analogy for the existence of God.

[2] Or more recently, following Owen (1960), 'focal meaning'.

cases where a word had a primary use but also secondary uses, where the secondary uses were explained by their relation to the primary use – thus in its primary use animals and plants are healthy, but a climate is also called healthy if it promotes health in plants and animals. Eventually it had been this second sense of *analogia* that tended to dominate the discussion rather than the idea of proportional analogy that concerns us. Thus, when we look at Aquinas, who is in many ways the most important figure in mediaeval discussions of analogy, in his mature writings he gives an account of the analogy between God and the world that relies not on Aristotelian proportional analogy, but on the analogy of attribution, relegating proportional analogy to its role in his account of metaphor.[3] So, when Cajetan in 1498 wrote a book giving a taxonomy of analogy[4], although he himself advocated proportional analogy, this was only one subdivision of '*analogia*'.

We therefore in this chapter move forward to the eighteenth century, to look at the treatment of analogy in the period leading up to the publication of the *Origin*. What we find in this period are two very different understandings of the word 'analogy', and as a consequence two very different understandings of what an argument by analogy is. A number of writers continue to explore analogy in its original Greek sense, leading eventually to Richard Whately's investigation of argument by analogy – when it succeeds and when it fails. This last is to this day one of the best treatments of the subject. Within this tradition, there is an Aristotelian insistence on the contrast between analogy as proportionality and simple similarity. However, it is even more popular in this period for writers to completely disregard this contrast and to treat 'A is analogous to B' as equivalent to 'There are properties shared by A and B', or possibly 'several properties'. Here a so-called argument by analogy is based on the idea that if A and B are known to share several properties, that makes it likely that some property of A that interests us is also shared by B.

The significance of this for our book is that the second account of argument by analogy became and remained highly popular and the overwhelming majority of authors who have discussed Darwin's use of analogy in the *Origin* assume this account of analogy. We shall be arguing in the following chapters that a careful reading of the text of the first four chapters of the *Origin* shows instead that Darwin is presenting a near

---

[3] For a fuller discussion, see Roger M. White (2010) Chapter 4. (Cf. also White (1982))
[4] Cajetan (1498).

perfect example of argument by analogy following the classic Greek pattern.

'Analogie' in the *Encyclopédie*

Hitherto, however the word 'analogy' has been understood, one thing was generally agreed: analogy is to be contrasted with simple similarity – the possession by two objects of a number of common properties. Beginning in the eighteenth century, a number of authors ignored this contrast – whether through ignorance of this history, or in reaction against it. This set the scene for the treatment of the word 'analogy' as a highly general term, applicable to a wide variety of different sorts of similarity. A representative of what would become the popular understanding of what was meant by 'analogy', and as a consequence by argument by analogy is to be found in the *Encyclopédie*.

The article on analogy in the first volume[5] of Diderot and d'Alembert's *Encyclopédie* was written by Urbain de Vandenesse. He was a physician who contributed a large number of articles to the *Encyclopédie*, of which the great majority were on medical topics. He cannot therefore be considered as an expert on the topic of analogy, but for that very reason he is ideal for our purposes. His article may be taken as a representative account of the *popular* understanding of the concept of analogy in the mid-eighteenth century.

The first thing to strike one is the almost complete disappearance of Aristotelian analogy from the whole article. Whereas among the schoolmen it was recognised as one important species of analogy, here there is no understanding that there was anything particular or important about the kind of four-term comparison envisaged by the Greeks. Instead for the most part 'analogy' is treated as a generalised notion of similarity.

When Vandenesse does begin with a brief paragraph concerning the classical sources, Aristotle and the Greek background is ignored, the only authority cited being Cicero:

> Cicero says that since he uses this word in Latin, he will translate it as comparison, *a relation of resemblance* between one thing and another.

When he then turns to the schoolmen, he begins, without citing any authority:

> Scholastics define analogy as a resemblance coupled with some diversity.

---

[5] Published in 1751. We are grateful to Emily Herring for help with the English translation.

It was, of course, one of the reasons for mediaeval theologians and philosophers interesting themselves in analogy that it was analogy that made possible comparisons between things that were different in kind.[6] This was, however, for them a consequence of the nature of the analogical relationship, not its definition. Vandenesse then sketches Cajetan's taxonomy of analogy, including the only explicit reference in the article to Aristotelian analogy:

> The final kind of analogy is proportional analogy, in which, although the reasons for using the same name are veritably different, they nonetheless share a relationship of proportionality: in this sense, the *gills* of fish are said to be *analogous* to the *lungs* of terrestrial animals. In addition, the eye and the faculty of understanding are said to be analogous or to relate to one another.

Against this background, Vandenesse's own definition of analogy is

> Analogy therefore designates the relation, the relationship or the proportion between several things which otherwise differ by their specific qualities. Thus, the base of a mountain might have something analogous to the feet of an animal, although these two things are very different.

Here he seems to agree with the account of analogy as similarity in dissimilarity. However, in what follows, and in particular in his examples of argument by analogy even this qualification is dropped, and he is prepared to countenance *any* similarity as an analogy.

His first example is, as he himself is fully aware, preposterous:

> Analogy is also one of the grounds of our reasoning; I mean that we are often put in the situation of producing explanations which prove nothing if they are only founded on analogy. For instance, there is a constellation in the sky called lion: the analogy between this word and the name of the animal that we name lion has given astrologists the impression that children born under this constellation are of a martial disposition: this is a mistake.

What is of interest to us is not, however, the ludicrous nature of the argument, but the fact that he regards it as an argument by analogy. It is utterly obscure what the 'form' of this 'argument' is supposed to be, and it looks as if any argument based in no matter how obscure a way upon supposed similarities is now treated as an argument by analogy.

---

[6] Compare the account of the analogy between God and the world from the Fourth Lateran Council (1215): 'For every similarity between Creator and creature, however great, one must always observe an ever-greater dissimilarity between them.'

Vandenesse now offers more reasonable examples of arguments, but here even the simplest inductive arguments are presented as arguments by analogy. Once we count the argument 'tree A is the same kind of tree as tree B. Therefore the fruit of tree A will taste the same as the fruit of tree B' as an argument by analogy, even the idea that analogy was differentiated from simple similarity by the qualification 'similarity *in dissimilarity*' has been dropped.

> Analogical reasoning can be used to explain or to clarify certain things, but it cannot demonstrate them. Nevertheless, a large part of our philosophy is founded upon analogy. It is useful for it saves us thousands of pointless discussions which we would otherwise have to repeat for each particular body. It suffices that we know that everything is governed by general and constant laws to have the grounds to believe that bodies which appear similar share the same properties, that fruit from a same tree have the same taste, etc.

The overall impression left by this article is one of confusion, and a lack of a clear sense of what the '*analogie*' actually means. Above all for our purposes, analogy in its original sense has almost completely disappeared, although the word 'proportion' does recur, the only recognition of its possible theoretical significance is restricted to two references – once as one of Cajetan's types of analogy, and then in a brief paragraph where it is restricted to its use in mathematics:

> ANALOGY in Mathematics is the same thing as proportion or equality of ratios.

To see what happens to the idea of an argument by analogy once the original understanding of the word ' analogy' has been lost, we turn from this popular account to an altogether more disciplined account – one that has been widely influential, even to the present day.

Thomas Reid's Account of Argument by Analogy

In *Rhetoric* Book II, Aristotle distinguishes two different types of 'example' (παράδειγμα):

> Let us start with example; for example resembles induction, and induction is a principle of reasoning. There are two sorts of example – those which refer to things which have actually happened, and those which use made up cases. The latter are subdivided into parables (παραβολή) and fables, such as those of Aesop and the Libyan fables. A case of the historical type of example would be to say that it is necessary to make preparations against the

great king and prevent him from conquering Egypt; for Darius did not invade Greece until he had first overcome Egypt, but immediately he had done so, he did. Similarly Xerxes did not attack us until he had taken Egypt, but once he had done so, he crossed over. Therefore, if the great king takes Egypt, he will cross over, and we must stop this happening. *Parable* is illustrated by the sayings of Socrates – e.g., saying that we ought not to choose our magistrates by lot, since this would be like choosing the athletes to represent us at the Olympic Games by lot rather than by their skill at athletics, or like sailors choosing their helmsman by lot, rather than one with the relevant knowledge.[7]

Aristotle is here separating out two superficially very similar types of argument: in both cases we have the presentation of what happens in one case and an inference from that case to what will happen in a parallel case. But despite appearances, Aristotle is right to separate them, since the two types of argument work entirely differently. We can contrast the two cases in a number of ways, including Aristotle's own point that, whereas the second argument can be equally effective with fictitious or hypothetical premises, the first argument essentially depends on the fact that Darius and Xerxes actually behaved as reported. The contrast that concerns us, however, is that it is only 'examples' of the second type that are in a strict sense arguments by analogy – for instance, in the case cited by Aristotle and that we referred to at the beginning of Chapter 1, the argument depends on the analogy 'magistrates are to public affairs as helmsmen are to ships': the examples of the first type appear to be simple examples of arguments by induction.

We began with the word 'analogy' being used exclusively by authors such as Aristotle for the four-term relation 'A is to B as C is to D'.[8] Although the schoolmen increased the scope of the word, they continued to recognise Aristotelian analogy as one important species of *analogia*. Now we are moving into a period in which the significance of analogy in its original sense has been lost, and as a result, the word 'analogy' came to be used most frequently as a synonym for similarity.

Against this background, when Thomas Reid gives an account of argument by analogy, he writes:

> We may observe a great similitude between this earth which we inhabit, and the other planets, Saturn, Jupiter, Mars, Venus, and Mercury. They all revolve around the sun, as the earth does, although at different distances

---

[7] *Rhet.* II, 1393a 22ff.
[8] Cf. e.g. *NE* V. 1131a: 'analogy is an equality of ratios, involving four terms'.

and in different periods. They borrow all their light from the sun, as the earth does. Several of them are known to revolve around their axis like the earth, and by that means, must they have a succession of day and night. Some of them have moons, that serve to give them light in the absence of the sun, as our moon does to us. They are all, in their motions, subject to the same law of gravitation, as the earth is. From all this similitude, it is not unreasonable to think that those planets may, like the earth, be the habitation of various orders of living creatures. There is some probability in this conclusion from analogy.[9]

This clearly conforms to Aristotle's first type of 'example',[10] and Aristotle's second type of 'example' – the only one that Aristotle himself would have acknowledged as argument by analogy – has disappeared. The argument runs as follows: we have two objects A and B, of which A is known to be F, and we want to know whether B is also F. A and B share a number of properties: G, H, I, etc. The more properties they share, the more probable it is that B is F, and every property shared increases the probability that B is F.

This account became highly influential and even today authors will present a variant of Reid's account as *the* explanation of what is meant by 'argument by analogy'. In the nineteenth century then we find two incompatible versions of what an argument by analogy was: one going back to the Greek original and the other derived from Reid.

As we shall see, one of the most judicious accounts of the situation is given by John Stuart Mill who, in the chapter in his *A System of Logic*, 'of Analogy'[11] presents both kinds of argument. Like Aristotle, he both presents the two kinds together and clearly differentiates them. The first, 'Aristotelian' kind, once it is established that there is an analogy, 'has the force of a rigorous induction'. The second 'Reidian' kind is then examined at length, the crucial sentence being

> If, however, every resemblance proved between B and A, in any point not known to be immaterial with respect to *m*, forms some additional reason for

---

[9] Reid (1785), Essay 1, ch. 4.
[10] Beginning towards the end of *Topics* I (108b 10ff.), there is a discussion of arguments based on similarity, which is very close to Reid's 'argument by analogy'. In particular, Aristotle says 'it is received opinion that if something is true of one of several things that are similar, it is also true of the rest' (108b, 15). There are two points to note about this: firstly that Aristotle seems here to be thinking of the rhetorical efficacy of such reasoning rather than its genuine probative power, and secondly that given his understanding of analogy, he would not have thought of this argument as having anything to do with argument by analogy, unlike the case of 'παράδειγμα' that we looked at in Chapter 1.
[11] Book III, ch. XX. On Mill's response to the *Origin*, see Hull (2009), pp. 190–193.

presuming that B has the attribute *m*; it is clear, *è contra*, that every dissimilarity which can be proved between them furnishes a counter-probability of the same nature on the other side.

Mill's is more sophisticated than Reid's original in two respects: firstly, in giving weight to dissimilarities as well as similarities, but secondly and most importantly, by the inclusion of the phrase 'in any point not known to be immaterial with respect to *m*'. There is no hint of this in Reid, but without it the whole account is preposterous. Any two objects have countless common properties that are totally irrelevant to the topic in hand. When we say that A is F, G, H and I, therefore B's being F, G and H increases the probability of its being I, the only properties F, G and H that matter are ones that have some bearing on the fact that A is I. Thus, people see the presence of water on Mars as increasing the probability of life on Mars, but the discovery of bauxite on Mars would have no such implication.

The introduction of the concept of *relevance* into the account implies that it is impossible to give a rigorous formal account of Reidian argument by analogy: what is relevant varies uncontrollably from case to case – the conquest of Egypt is relevant to the invasion of Greece in a completely different way from the presence of water on Earth to the presence of life. It will also inevitably introduce some subjectivity into the decision as to what is and what is not relevant.

We may highlight the following points concerning Reid's account. We have here an account in which analogy in its original sense (A:B::C:D) plays no part whatsoever. Whereas, properly conducted, an Aristotelian argument by analogy is a valid argument,[12] Reid's argument is at best a probabilistic argument. Although Reid's account became increasingly popular, the original Aristotelian account was also championed in the nineteenth century.[13] Reid's account leads nowhere in understanding the use of analogy in the *Origin*, where it is only the original 'Greek' argument by analogy that helps us understand Darwin's procedure.

William Paley, *Natural Theology*

Because of its role in discussions of the significance of Darwin's theory of natural selection nowadays, we need to include William Paley's *Natural*

---

[12] Whether it is also a *sound* argument will of course depend on answering further questions, such as the question whether there is a genuine analogy in a particular case.
[13] Above all by Richard Whately.

Theology. This was published in 1802. It was not a highly original book,[14] presenting as it did an argument for the existence of God based upon evidences of design in the world and in particular in the animal kingdom that had been common in post-Enlightenment Christian Apologetics. Its strength lay in the clarity of its exposition, together with the way that Paley confronts readers with the vast range and subtle detail of the apparent 'evidences of design' in the animal kingdom. It is now almost universally regarded as having been radically undermined by the *Origin of Species* which presents us with an alternative, purely naturalistic explanation of the phenomena that for Paley could only be accounted for by positing a benign God. However, precisely for this reason it is easy to misconceive the relation between the two books.

> Although I did not think much about the existence of a personal God until a considerably later period of my life, I will here give the vague conclusions to which I have been driven. The old argument of design in nature, as given by Paley, which formerly seemed to me so conclusive, fails, now that the law of natural selection has been discovered.[15]

Although the work of Paley's that most impressed Darwin was his *Evidences of Christianity*,[16] Darwin also thought highly of his *Natural Theology*. For most people nowadays, their conception of and interest in Paley's work will be restricted to the idea that he presented an argument for the existence of God based on an analogy between the relation of God to the world and a watchmaker and a watch, an argument that was persuasive until it was delivered a death blow by the argument of the *Origin*. Taken together, these two facts may easily lead to an exaggerated idea of the influence of Paley on Darwin's thought, and a distorted account of Paley's argument and its relation to Darwin's alternative explanation of the existence of animals and plants. What we shall do here is attempt first to dispel some misconceptions, and then single out one feature of *Natural Theology* that may well have influenced Darwin's development of the argument of the *Origin* – a feature that throws light on an important aspect of analogical reasoning.

We may specify three possible misconceptions that are to be found in discussions of *Natural Theology* and its relation to the *Origin*: that Darwin was led to the theory of evolution by natural selection by a quest for a naturalistic alternative to Paley's theistic account of the origin of animal

---

[14] Quite soon, Paley was accused of plagiarism. There are indeed striking similarities between *Natural Theology* and Nieuwentyt (1721). But for our purposes, it was of course Paley that Darwin read.
[15] Darwin (1958), p. 87.   [16] Paley (1794).

and plant life; that the argument of *Natural Theology* is an argument by analogy; and that the argument of the *Origin* is a secular alternative to Paley's argument with a structure paralleling that of Paley but with natural selection replacing God in the conclusion.

> He (Darwin) found it (*Natural Theology*) one of the few stimulating books he had to read there (at Cambridge) and wrestled with finding an alternative to Paley's vision of design. He found it in the hidden hand of natural selection: the survival of the fittest in the struggle for existence.[17]

Thus the editors of the Oxford World's Classics. It suggests a very different account of what led Darwin to the theory of evolution by natural selection from that implied by the passage just quoted from Darwin's autobiography. There is no reason whatever to think that replying to Paley had any part in the discovery of his theory. He was far more concerned to account for the manifold evidences for some form of evolution having occurred, such as the geographical distribution of species and the fossil record. As he himself says, it was with retrospect that he realised that his theory constituted a rebuttal of Paley. Because of this, as we shall see, Darwin's argument in the *Origin* will have a very different form from that of *Natural Theology*, and not simply be an argument with the same structure but proposing a naturalistic alternative to Paley's theistic explanation of the phenomena.

But first we shall look at Paley's argument in its own right, asking whether, as is usually assumed, it really is an argument by analogy. What creates the impression that it is such is the famous opening sentences:

> In crossing a heath, suppose I pitched my foot against a *stone*, and were asked how the stone came to be there; I might possibly answer, that, for any thing I knew to the contrary, it had lain there for ever: nor would it perhaps be very easy to show the absurdity of this answer. But suppose I had found a *watch* upon the ground, and it should be inquired how the watch happened to be in that place; I should hardly think of the answer which I had before given, that, for any thing I knew, the watch might have always been there. Yet why should not this answer serve for the watch as well as for the stone? Why is it not as admissible in the second case, as in the first? For this reason, and for no other, viz. that, when we come to inspect the watch, we perceive (what we could not discover in the stone) that its several parts are framed and put together for a purpose, *e.g.* that they are so formed and adjusted as to produce motion, and that motion so regulated as to point out the hour of the day; that, if the different parts had been differently shaped

---

[17] Paley (1802) (eds. Matthew D. Eddy & David Knight 2006. Editors' Preface, p. xxvii).

from what they are, of a different size from what they are, or placed after any other manner, or in any other order, than that in which they are placed, either no motion at all would have been carried on in the machine, or none which would have answered the use that is now served by it.[18]

This, initially at least, suggests that Paley is going to argue by analogy, at least in Reid's sense of argument by analogy, along the following lines: we know from experience that artefacts such as watches, which are complexes of parts arranged in such a way as to produce a functioning whole are only ever the product of intelligent beings. However, human beings together with the rest of the animal kingdom are such complexes. Therefore it is probable that they too are produced by an intelligent being.

However, as Paley continues, he makes it clear that it is not his intention that he should be understood in this way:

> Nor would it, I apprehend, weaken the conclusion, that we had never seen a watch made; that we had never known an artist capable of making one; that we were altogether incapable of executing such a piece of workmanship ourselves, or of understanding in what manner it was performed; all this being no more than what is true of some exquisite remains of ancient art, of some lost arts, and, to the generality of mankind, of the more curious productions of modern manufacture.[19]

That is to say, he does not wish his argument to depend on our familiarity with watches and their manufacture. But if he is not arguing by analogy from familiar facts about watches, what is the point of the reference to the watch? The answer is that we have here an instance of what William Kneale called 'intuitive induction',[20] where the induction 'exhibits the universal as implicit in the clearly known particular'.[21] That is to say, the use of the example of the watch is to introduce us to a general idea or principle whose application is clearly exhibited by this particular case.[22] This way of reading Paley is confirmed by the following passage:

> Were there no example in the world, of contrivance, except that of the *eye*, it would be alone sufficient to support the conclusion which we draw from

[18] Ibid., p. 7.   [19] Ibid., p. 8.

[20] *Cf.* the discussion in Kneale (1948), pp. 30–37. 'Intuitive induction' is Kneale's translation for one of Aristotle's uses of the term "ἐπᾰγωγή".

[21] *PostA*, 71a 8.

[22] Two authors, who, like us, reject the idea that Paley is offering us an argument by analogy, are Sober (2003), and Jantzen (2014). For Sober, the argument is an argument to the best explanation, for Jantzen, it is an argument with the premise 'Goal-directed systems can only be produced by design' (Jantzen, 2014, p. 133). One or other of these might give us formulations of the general principle that is for Paley illustrated by the watch.

it, as to the necessity of an intelligent Creator. It could never be got rid of; because it could not be accounted for by any other supposition, which did not contradict all the principles we possess of knowledge.[23]

That is to say, the function of the example of the watch is simply to illustrate a 'principle of knowledge' – something along the lines that confronted by a complex entity that displays purposiveness, we refuse to allow that it could be the product of chance, but instead are forced to the conclusion that there is an intelligence responsible for the entity. This 'principle' enables us to infer that given that the eye is such a complex entity, there is an intelligence responsible for it.

What may lead to the impression that we are here dealing with an argument by analogy is that there is analogy, in the strict Aristotelian sense, involved. Namely, the analogy God is to the human eye as a watchmaker is to a watch. However, that is the *conclusion* of the argument, not its premise. What we have here is an argument *to* analogy, not an argument *from* or *by* analogy.

Because Darwin is thought to have provided a complete rebuttal of the argument of *Natural Theology*, it is natural to assume that the argument of the *Origin* will mimic Paley's argument with natural selection replacing God in the conclusion. But in fact the argument of the *Origin* proceeds in almost the exactly opposite way to Paley's. Whereas, as we have just said, Paley's argument may be regarded as an argument *to* analogy, Darwin's is an argument *from* analogy. If Darwin's argument paralleled Paley's its conclusion would be the need to posit a struggle for existence in order to account for the emergence of new species. But Darwin uses the idea of a Malthusian struggle for existence to argue directly for the existence of natural selection, in order thereby to establish the analogy: natural selection is to animals in the wild as artificial selection is to animals under domestication. This will then form the basis for an argument to the conclusion that just as artificial selection produces new varieties under domestication so natural selection will produce new varieties in the wild. This is an argument that proceeds in almost the exactly opposite direction to Paley's, with Darwin's premises corresponding to Paley's conclusion and *vice versa*.

So far we have argued for a negative account of the relation of Paley's work and the *Origin*. But there is one highly instructive parallel between their procedures, and where indeed Darwin might well have learnt from

---

[23] Paley (1802), p. 45.

Paley and been influenced by him. Paley describes his argument as 'cumulative in the fullest sense of that term'.[24] It is unclear how he means this, since he continues, 'The eye proves it without the ear; the ear without the eye'. So the question arises, 'Why repeat essentially the same argument again and again at such tedious length?'[25] If Paley wished only to establish the analogy 'God is to the eye as a watchmaker is to a watch', there seems no satisfactory answer to that question. But Paley is concerned to establish a stronger conclusion than that simple analogy. The formula thus far does nothing to indicate that God is greater than any human being. What will show that God *is* greater is the superiority of God's works to anything that could be achieved by a human being. It is this that explains the lengthy survey of all the complex and intricate detail of a human being showing something constructed with a subtlety that is far beyond the reach of any human artefact. From this we can simply conclude that the superiority of the workmanship shows the superiority of the maker.

> It is an immense conclusion, that there is a GOD; a perceiving, intelligent, designing, Being; at the head of creation, and from whose will it proceeded. The *attributes* of such a Being, suppose his reality to be proved, must be adequate to the magnitude, extent, and multiplicity of his operations: which are not only vast beyond comparison with those performed by any other power, but, so far as respects our conceptions of them, infinite, because they are unlimited on all sides.[26]

It is doubtful whether Paley is really entitled to his 'infinite' here. But what interests us is the step that Paley takes. In effect what he is doing is taking the analogy 'God is to an animal as a watchmaker is to a watch', say, and alternating it: 'God is to a watchmaker as an animal is to a watch'. That is to say, the better the product, the better the producer.

It is precisely this manoeuvre of alternating the analogy that Darwin will exploit in Chapter IV of the *Origin*. There his problem is to show that it is at least possible that natural selection should achieve something that artificial selection apparently never did, namely produce the kind of irreversible change constituted by the emergence of a new species. For that, he argues that natural selection is a vastly more powerful selector than any human breeder, and hence that it is possible that its effect should be

---

[24] Ibid., p. 46.
[25] Eddy (2004) cites Cicero, *De Natura Deorum*: 'And if perchance these arguments separately fail to convince you, nevertheless, in combination their collective weight will be bound to do so.' The comparison is indeed striking, but that alone doesn't seem to justify the book length repetition of the same argument.
[26] Eddy (2004), p. 230.

vastly greater. This is essentially the same move as the one we have just seen: natural selection is to modifications of creatures in the wild as artificial selection is to modifications of creatures under domestication. Therefore natural selection is to artificial selection as modifications of creatures in the wild is to modifications of creatures under domestication – the more powerful the selector, the more powerful its effect. If Darwin learnt anything from Paley, this is what it may have been. Though even here his application once again proceeds in the opposite direction to Paley's: where Paley had argued 'The greater the product, the greater the producer', Darwin will argue 'The greater the producer, the greater the product.'

Aristotelian Analogy in the Eighteenth and Nineteenth Centuries

> Every one knows, that analogy is a Greek word used by mathematicians, to signify a similitude of proportions. For instance, when we observe that two is to six as three is to nine, this similitude or equality of proportion is termed analogy. And although proportion strictly signifies the habitude or relation of one quantity to another, yet, in a looser and translated sense, it hath been applied to signify every other habitude; and consequently the term analogy comes to signify all similitude of relations or habitudes whatsoever.[27]

Although the idea that analogy could simply be regarded as synonymous with similarity was widespread throughout the eighteenth and nineteenth centuries, there were still a number of writers who were conscious of its original meaning and insisted that this was the correct meaning to assign to the term. In particular, there was an intense debate over the nature of the analogy between God and the world among orthodox Anglican theologians, in which despite theological disagreements there was shared understanding of what was meant by the word 'analogy'. We are obviously not concerned here with the theological issues, but only with what happens to the concept of analogy itself in this debate.

In 1709, William King[28] preached a sermon, 'Predestination and Foreknowledge Consistent with the Freedom of Man's Will'. In it, he argues that, given the radical difference between God and His creation, when we talk about God using our human language we are forced to resort to metaphors and analogies. Although in the context of the sermon he does

---

[27] Berkeley (1732), Dialogue IV, section 21.
[28] William King, 1650–1729. Archbishop of Dublin.

not clarify how he understands the term 'analogy', his illustrations make best sense if we interpret him as using analogical models (using the idea of analogy in its original Greek sense):

> Thus when we would help a Man to some Conception of any thing, that has not fallen within the reach of his Senses, we do it by comparing it to something that already has, by offering him some Similitude, Resemblance or Analogy, to help his Conception. As for Example, to give a Man a Notion of a Country, to which he is a Stranger, and to make him apprehend its Bounds and Situation, we produce a Map to him; and by that he obtains as much knowledge of it as serves him for his present purpose. Now a Map is only Paper and Ink, diversify'd with several Strokes and Lines, which in themselves have very little likeness to Earth, Mountains, Valleys, Lakes and Rivers. Yet none can deny, but by Proportion and Analogy they are very instructive; and if any should imagine that those Countrys are really Paper, because the Maps that represent them are made of it, and should seriously draw Conclusions from that Supposition, he would expose his Understanding, and make himself ridiculous: And yet such as argue from the faint Resemblances, that either Scripture or Reason give of the Divine Attributes and Operations, and proceed in their Reasonings, as if these must in all respects answer one another, fall into the same Absurdities that those would be guilty of, who should think Countrys must be made of Paper, because the Maps that represent them are so.

King's example of a country and a map of that country gives a clear case of the kind of analogical model that interests us. This means that anyone who studies a map and as a result says 'To go from town A to town B, we must cross a river' is making a sound inference by analogy. (With, of course, the tacit premise that the map has been properly constructed – by triangulation, say.) King then can be seen as arguing that the doctrines of the Christian faith can be best understood as describing analogical models of the divine reality. The appearance then that there is conflict between different doctrines, such as the doctrine of predestination and freewill is the result of having pressed one or more of these analogies too far.

This sermon was extraordinarily influential and provoked a long debate into the nature of the analogy between God and His creation. Key figures in that debate were Peter Browne,[29] George Berkeley, Edward Copleston and Richard Whately. Several ideas recur throughout the debate that are relevant to our enquiry. Analogy could only here be understood in its original strict sense of a four-term relation, 'A is to B as C is to D'; this was

---

[29] Browne (1733). Peter Browne (1665–1735) was Bishop of Cork and Ross.

to be clearly distinguished from the widespread 'loose' use of the term as meaning any sort of similarity; analogy in the strict sense, and only analogy in the strict sense, enables us to make significant comparisons between completely heterogeneous entities; and extreme care must be taken not simply to assume that because two things were analogues they must share yet further properties.

We shall look at Copleston and Whately later, but for now we shall look briefly at the fierce dispute between Browne and Berkeley. Browne's position, developed under the influence of King's sermon, was a version of the position that is worked out with full rigour by Kant, and which we shall look at next. According to this, the analogy between God and the world was such as only to permit us to talk about the way in which God related to the world, and told us nothing about the properties of God as He is in Himself. What God was actually like in Himself was inconceivable.

For Berkeley such a position was indistinguishable from atheism:

> For someone who comes to God, or goes into the church of God, must first believe that there is a God in some intelligible sense; not merely that there is something in general, without any proper notion – even a very inadequate one – of any of its qualities or attributes; for this 'something in general' could be fate, or chaos, or creative nature, or anything else, as well as it could be God. And it's no help to say there is something in this unknown being that is analogous to knowledge and goodness; i.e. something that produces the effects that we can't conceive to be produced by men without knowledge and goodness. For this is still to surrender to the atheist side against the theists. ... You cannot argue from unknown attributes, or which is the same thing, from attributes in an unknown sense. You cannot prove, that God is to be loved for his goodness, or feared for his justice, or respected for his knowledge: all which consequences, we own, would follow from those attributes admitted in an intelligible sense; but we deny that those or any other consequences can be drawn from attributes admitted in no particular sense, or in a sense which none of us understand[30]

He then goes on to give his own account of the analogy between God and humanity, an account that Browne saw as failing to do justice to the radical difference between God and humanity:

> Knowledge, therefore, in the proper formal meaning of the word, may be attributed to God proportionably, that is, preserving a proportion to the

---

[30] Berkeley (1732), Dialogue IV, Section 18.

infinite nature of God. We may say, therefore, that as God is infinitely above man, so is the knowledge of God infinitely above the knowledge of man, and this is what Cajetan calls *analogia proprie facta.*[31]

With a little charity on both sides, it would be possible to reconcile Browne and Berkeley's apparently opposed positions. The relevance of this debate to our concerns is that when we move away from this theological debate to Darwin's use of analogy, we shall find that Darwin is in effect applying analogy both in Browne's way (artificial selection is to animals under domestication as natural selection is to animals in the wild) and in Berkeley's way (as natural selection is far above artificial selection, so are the effects of natural selection far above the effects of artificial selection). We shall therefore return to the question of the relation between these two apparently divergent applications of analogy.

Immanuel Kant: Strictures on the Use of Analogy

This type of cognition is cognition *according to analogy*, which surely does not signify, as the word is usually taken, an imperfect similarity between two things, but rather a perfect similarity between two relations in wholly dissimilar things.[32]

Darwin's knowledge of Kant's work was extremely limited, so that it is unlikely that he would be aware of Kant's treatment of analogy in particular. However, we include Kant in our survey of the background to analogy in the *Origin*, since he gives the most rigorous treatment of analogy in this period, showing the severe conditions that must be met for a sound argument by analogy.

He discusses analogy in a variety of contexts, but the two most important ones with which we shall be concerned are in the introductory section to 'The Analogies of Experience' in the *Critique of Pure Reason* and in his account of the use of analogy to explain religious language in the *Prolegomena*. Two interrelated ideas run through all his discussions: a contrast between the use of analogy within mathematics and outside mathematics and the fact that analogy enables us to compare completely heterogeneous entities that otherwise share no common properties.

In *The Analogies of Experience,* Kant is concerned to establish his account of what can be called the uniformity of nature. In his introduction to this section of the first *Critique*, he explains this as a matter of

---

[31] Ibid., Dialogue IV, section 21.   [32] Kant (1783), pp. 146–147 (Ak. 4:357).

establishing systematic analogical relationships between what happens at different points of space/time. 'The principles can ... have no other purpose save that of being the conditions of the unity of *empirical* knowledge in the synthesis of *appearances*.'[33]

We are not concerned here with the complexities of the detailed arguments that follow, or the question whether or not Kant succeeds in justifying each of his three analogies. What does concern us is the idea running through his discussion that we can put in the form that such an analogical structure gives us the form of any possible scientific theory but not its content, so that, for example, *The First Analogy* will establish a Principle of Conservation, but what it is that is conserved will necessarily be a matter of empirical investigation. This claim is to be justified by his contrast between the mathematical and the non-mathematical employment of analogy.

> In philosophy analogies signify something very different from what they represent in mathematics. In the latter they are formulae which express the equality of two quantitative magnitudes, and are always *constitutive*, so that, if three terms of the proportion are given, the fourth is thereby given, that is, can be constructed. But in philosophy the analogy is not the equality of two quantitative but of two qualitative relationships, and from three terms we can acquire *a priori* knowledge only of the relation to a fourth, not of the fourth term itself. The relation yields, however, a rule for seeking the fourth term in experience, and a mark whereby it can be detected. An analogy of experience is, therefore, only a rule according to which a unity of experience may arise from perception. It does not tell us how mere perception or empirical intuition in general itself comes about. It is not a principle *constitutive* of the objects, that is, of the appearances, but only *regulative*.[34]

If we consider a simple mathematical analogy, $51/17 = x/114$, it is a matter of simple arithmetic to compute that $x = 342$. That is to say, given three terms of the analogy, what the fourth term is is guaranteed a priori. Similarly in the case of the more complicated application of analogy within mathematics – that of similar triangles that we looked at in Chapter 1 – given any two similar triangles, we can provide geometrical proofs showing which properties are necessarily shared by the two triangles. What this means is that within mathematics the idea of arguments by analogy that are not only valid but also sound is completely unproblematic.

However, when we look at applications of analogy outside mathematics, there is characteristically no longer any way of determining a priori what

[33] Kant (1787), B224.   [34] Ibid., B222–B223.

the fourth term of the analogy and what its properties are. 'The *existence* of appearances cannot, however, be thus known *a priori*; and even granting that we could in any such manner contrive to infer that something exists, we could not know it determinately, could not, that is, anticipate the features through which its empirical intuition is distinguished from other intuitions.'[35] Kant therefore assigns to the use of analogy outside mathematics a purely 'regulative' function – 'a rule for seeking the fourth term in experience, and a mark whereby it can be detected'. We can see what for Kant is the legitimate use of analogy, if we return to the use Aristotle makes of it in his biology. Having, for example, seen that elephants can grasp things, Aristotle is justified in claiming that there will be a part of the elephant that is the analogue of the human hand. But it then requires empirical observation of the elephant to discover that it is the trunk that performs this function, and also to discover what features the trunk shares with the hand to enable it to do so.

If, however, such empirical investigation is in the nature of the case impossible, then Kant claims that it is impossible to infer from the existence of the analogy anything about the intrinsic properties of the analogues:

> [W]hen we compare the artful acts of animals with those of man, we do not know what basis in these animals gives rise to such effects, but we do know what basis gives rise to similar effects in the case of man (namely, reason); and hence we conceive of the basis for such acts in animals by means of the basis of such acts in man: *i.e.*, we conceive of the former basis as an analogue of reason. In doing so we wish to indicate at the same time that the basis of the artistic power in animals, called instinct, while indeed different in kind from reason, still has a similar relation to its effect (for example, if we compare the constructions of beavers with those of human beings). But that does not entitle me to infer that because man needs *reason* in order to construct things, beavers too must have it, and call this an *inference* by analogy.[36]

In the same way, Kant in his philosophy of religion in the *Prolegomena* will see the assertion of an analogy between God and the world as telling us nothing about God's intrinsic properties, restricting us to knowing only about the *relation* of God to the world. (He would have been appalled by the easy way in which Paley ascribes human characteristics to God.)

This train of thought places severe restrictions on the use of analogy, limiting its use to a guide to future empirical enquiry, and thus apparently

---

[35] Ibid., B221.   [36] Kant (1793), pp. 356–357 (Ak. 5:464).

making a valid argument by analogy outside mathematics impossible. Kant himself may well have believed that to be so. However, what it *does* do is set stringent conditions that must be satisfied for such an argument to be possible. As we shall see, Darwin's own use of argument by analogy only requires appeal to something that Kant *will* allow, namely a doctrine of the uniformity of nature – the idea that the same pattern of causation will have the same kind of effect. What this means is that his argument will license the conclusion that natural selection will produce new varieties, and even possibly new species. It will not, however, give of itself any guide to which direction evolutionary change will take, or the nature of the new varieties that will result.

Edward Copleston, Richard Whately and the Aristotelian Revival

We are here mainly concerned with the treatment of analogy in Richard Whately's *Elements of Rhetoric*, the first edition of which was published in 1828. Nevertheless it is appropriate to include mention of Edward Copleston as a crucial part of Whately's background. As Whately freely acknowledged he was deeply indebted to Copleston, and his work can be regarded as both a presentation and an elaboration of ideas he had learnt from Copleston when he was tutored by him at Oriel College, Oxford.

Their major achievement was undoubtedly their restoration of logic as a serious academic discipline. This was not a matter of their making new technical advances in logic[37] but of a rescuing logic from the widespread misunderstanding and disparagement which followed above all the work of Locke. Two points stand out here. First, logic is purged of irrelevant psychological and epistemological ideas that, beginning with Locke, had been introduced into logic. For Copleston and Whately, logic was purely concerned with the logical form of propositions and arguments, and its aim was not to study how people in fact reason, but how they ought to reason. Second, there is a reply to a view that had become popular, that logic was pointless and logically valid arguments 'circular' – since all the information necessary to decide the truth of the conclusion was already contained in the premises, a valid argument told one nothing that one did not know already. The reply is that it would only be true that logic told us nothing 'new' if all the logical consequences of our beliefs were

---

[37] Indeed, Whately claimed that all valid arguments could be reduced to a series of syllogisms in Barbara. With hindsight, such a limiting conception would of itself stand in the way of serious advances in formal logic.

immediately clear to us. The function of logic should be to organise our ideas, and to make clear to us all the implications of what we knew.

Whately's *Elements of Logic* was widely influential and marked the beginning of the revival of logic in the United Kingdom. More significantly from the point of view of the present study, it also marked the beginning of an interest in Aristotle more generally. The treatment of analogy by Copleston and Whately can be seen as a bringing together of the ideas that we have been looking at in the discussion of the Archbishop King sermon and ideas already to be found in Aristotle. Copleston gave a commentary on King's sermon as an appendix to his set of discourses, 'An Enquiry into the Doctrines of Necessity and Predestination',[38] and Whately appended an annotated edition of the sermon to his Bampton Lectures, 'On the Use and Abuse of Party Feeling in Matters of Religion'.[39] Both were influential figures in the 'broad church movement' that permitted diversity in doctrine within the church, and in both cases, their concern with King's sermon was that it provided an argument for religious tolerance. The argument was that once we appreciated that it was only possible to have analogical knowledge of God, we would see that divisions within the church were caused by insisting on particular interpretations of the doctrines of the church that went beyond what could be justified by the analogy. To quote a typical passage from Whately:

> [S]ince our language when treating of heavenly things must be borrowed by analogy from things more level to our capacity, and since these analogies cannot but be very imperfect, the constant employment of the *same* analogical relations in each case respectively, will be apt to suggest to the hearer and fix in his mind some incorrect theory on the subject, by leading him to suppose the analogy more complete than in fact it is. The obvious prevention of the evil is to *vary* as much as possible the analogies made use of, that one may serve to correct the erroneous notions that might be suggested by another.[40]

Of most relevance to Darwin's use of analogy, however, is Whately's *Elements of Rhetoric*, of which the first edition was published in 1828. Before looking at some of the specific detail of Whately's treatment of analogy,[41] it is worth making some general points about this book. In the first place, it received a wide circulation, so that its ideas gained general currency. At the same time, if we look at the discussion so far, it would be possible to gain the impression that the discussion of analogy was a

---

[38] London, 1821. The commentary is an appendix to Discourse III.   [39] Delivered in 1822.
[40] Whately (1822), pp. 201–202.   [41] This is to be found in Whately (1828) Part I, ch. II.§7.

specifically theological topic, mainly of relevance in the philosophy of religion. However, although Whately's book was based on lectures for the training of ordinands, the great majority of his examples are drawn from the secular sphere, showing the pervasive character of analogy and analogical reasoning throughout the natural world. Finally, one of the merits of its treatment of analogy is its exploration of a wide range of examples.

When we look at the specific detail of the exploration of analogy, in part, in line with much of what we have seen so far, there are warnings against the illegitimate use of argument by analogy: 'In this kind of argument, one error, which is very common, and which is to be sedulously avoided, is that of concluding the *things* in question to be *alike*, because they are *Analogous*.' However, it is two other aspects of Whately's discussion that are directly relevant to Darwin's use of analogy in the *Origin*. In the first place, in the case of two arguments that he looks at, he at least indicates the conditions under which such arguments would be sound.

The first such argument is as follows:

> [S]uppose anyone had, at the opening of the French Revolution, or any other similar conjuncture, expressed apprehensions, grounded on a review of history, of the danger of anarchy, bloodshed, destruction of social order, general corruption of morals, and the long train of horrors so vividly depicted by Thucydides as resulting from civil discord, especially in his account of the sedition at Corcyra.

Whately first considers the reply of an opponent who points to the deep differences – political, cultural and technological, between the situation at the time of Corcyra, and that now obtaining, undermining this use of analogy. Thucydides himself had already anticipated such a reply, and Whately presents his rejoinder: 'he contends, that "as *long as human nature remains the same*", like causes will come into play, and produce, substantially like effects'. Thus, the argument is sound if and only if, as Thucydides contends, human nature is what John Stuart Mill will call the 'material circumstance': that is to say, it is human nature and human nature alone that causes the outcome of civil strife to be disastrous.

The other case in which Whately indicates the condition that would need to be satisfied for an argument by analogy to be sound is in his consideration of a supposed argument against protectionism: since it would be irrational for an individual to buy certain goods when other cheaper and superior goods were available, it must be irrational for a nation to buy goods produced at home when cheaper and superior goods are available from abroad. His final comment on this argument is: 'the question *important to the argument*, is, does the expediency, in private life, of obtaining each commodity at the least cost, and of the best quality we can, *depend* on any of the circumstances in which an Individual differs from a Community?' That is to say, the argument would be sound if, and only if, it could be shown that there were no relevant differences between what constituted financial prudence for individuals and communities.

In the second place, the other aspect of Whately's discussion that is directly relevant to Darwin's argument is that he spells out explicitly something that we have already mentioned as implicitly involved in Paley's argument, namely the possibility of a kind of a fortiori argument based on analogy:

> [I]llustrations drawn from things considerably remote from what is being illustrated will often have the effect of an '*a fortiori*' argument: as in some of the Parables . . . and that where Jesus says, 'If ye then, being evil, know how to give good gifts to your children, *how much more*', &c.
> So also in the Apostle Paul's illustration from the Isthmian and other Games: 'Now they do it to obtain a corruptible crown; but we an incorruptible.'

We shall look at this in detail when we look at Darwin's application of this idea, and at this stage just note a possible ambiguity in the idea of an a fortiori argument – an ambiguity that neither Whately or Paul resolve. The Pauline reference is to I Cor. 9, 24–25: 'Know you not that they who run in a race all run, but one receives the prize? So run, that you may obtain. And every man that strives for self control is temperate in all things. Now they do it to obtain a corruptible crown; but we an incorruptible.' We can paraphrase this compressed argument in two different ways, each of which can be described as an 'a fortiori' argument: *either,*

> Athletes competing for a prize exercise self-discipline. They only do so for a corruptible crown. But we are seeking an incorruptible crown. Therefore, we have even more reason to exercise self-discipline.

Or,

> Athletes competing for a prize exercise self-discipline. They only do so for a corruptible crown. But we are seeking an incorruptible crown. Therefore, we have reason to exercise even more self-discipline.

It seems impossible to determine which of these two readings was intended by Paul (or, for that matter, Whately). Of these two, it is the second way of understanding an 'a fortiori' argument that will be relevant when we come to discuss Darwin's use of analogy in the *Origin*.

John Stuart Mill, *A System of Logic*

For our purposes, the major interest in the chapter on analogy in Mill's *A System of Logic*[42] is that, being published in 1843, it provides an excellent and clear account of the state of the art in the years immediately preceding *The Origin of Species*. It divides naturally in two parts, the first, much briefer, paragraph dealing with Aristotelian analogy and the second longer section being introduced by saying 'It is on the whole more usual, however, to extend the name of analogical evidence to arguments from any sort of resemblance.' We have already looked at Mill's treatment of this second type of 'argument by analogy' in our discussion of Thomas Reid, and it is in any case the opening paragraph that is directly relevant for an understanding of Darwin's use of analogy. It is brief enough to quote in full.

§ 1. The word Analogy, as the name of a mode of reasoning, is generally taken for some kind of argument supposed to be of an inductive nature, but not amounting to a complete induction. There is no word, however, which is used more loosely, or in a greater variety of senses, than Analogy. It sometimes stands for arguments which may be examples of the most rigorous induction. Archbishop Whately, for instance, following Ferguson[43] and other writers, defines Analogy conformably to its primitive acceptation, that which was given to it by mathematicians: Resemblance of Relations. In this sense, when a country which has sent out colonies is termed the mother country, the expression is analogical, signifying that the colonies of a country stand in the same *relation* to her in which children stand to their parents. And if any inference be drawn from this resemblance of relations, as, for instance, that obedience or affection is due from colonies to the mother country, this is called reasoning by analogy. Or, if it be argued that a nation is most beneficially governed by an assembly elected by the people, from the admitted fact that other associations for a common purpose, such as joint-stock companies, are best managed by a committee chosen by the parties interested; this, too, is an argument from analogy in the preceding sense, because its foundation is, not that a nation is like a joint-stock company, or Parliament like a board of directors, but that Parliament stands in the same *relation* to the nation in which a board of directors stands to a joint-stock company. Now, in an argument of this nature, there is no inherent inferiority of conclusiveness. Like other arguments from resemblance, it may amount to nothing, or it may be a perfect and conclusive induction. The circumstance in which the two cases resemble, may be capable of being shown to be the *material* circumstance; to be

---

[42] Book III, ch. XX.   [43] The reference is to Ferguson (1792).

that on which all the consequences, necessary to be taken into account in the particular discussion, depend. In the example last given, the resemblance is one of relation; the *fundamentum relationis* being the management, by a few persons, of affairs in which a much greater number are interested along with them. Now, some may contend that this circumstance which is common to the two cases, and the various consequences which follow from it, have the chief share in determining all the effects which make up what we term good or bad administration. If they can establish this, their argument has the force of a rigorous induction; if they can not, they are said to have failed in proving the analogy between the two cases; a mode of speech which implies that when the analogy can be proved, the argument founded on it can not be resisted.

The only major quarrel that we have with this passage is that at crucial points it is much too compressed and demands considerable elaboration. We may first note the following observations about this passage. Mill is as clear as the authors that we have just been looking at that there are two different ways of understanding the word 'analogy', its original uses and a popular, broad, use. He claims that there is a type of argument by analogy, with analogy used in its strict sense, that is radically different from the popular type of argument derived from Reid. The former argument if properly conducted can be valid ('examples of the most rigorous induction'), whereas the latter are at best probabilistic arguments – the premises increasing the probability of the conclusion. The difference between the two arguments that Mill is identifying is not merely that one is based on relational properties, and the other on intrinsic properties. As Mill will point out later, an argument by analogy in Reid's sense can quite well be based on two objects having several relational properties in common. The difference is that there is a type of argument based on Aristotelian analogy that has no parallel argument based on two objects possessing common intrinsic properties. It is at this point that Mill's discussion demands further spelling out, particularly since, of all the authors we have considered in this chapter, Mill not only recognises the difference between arguments based on Aristotelian analogy and Reid's version of arguments by analogy but also indicates the conditions that need to be satisfied for an Aristotelian argument to be sound and fully conclusive. This is in the sentences: 'Now, some may contend that this circumstance which is common to the two cases, and the various consequences which follow from it, have the chief share in determining all the effects which make up what we term good or bad administration. If they can establish this, their argument has the force of a rigorous induction.' This is excellent, but needs unpacking. We shall return to this later, but for now we may briefly indicate Mill's thought here.

We are considering the argument 'Joint-stock companies are best managed by a committee chosen by the parties interested. ∴. A nation is most beneficially governed by an assembly elected by the people.' We first spell out the analogy between a joint-stock company and a nation ('the *fundamentum relationis* being the management, by a few persons, of affairs in which a much greater number are interested along with them'). We next consider the feature of joint-stock companies, that they are best managed by a committee chosen by the parties interested, and ask under what conditions the analogy permits us to transfer this feature to a nation. It is this question that is most frequently neglected in faulty arguments by analogy. Mill's answer, which clearly makes sense, is that the feature can be transferred if and only if it can be shown that a joint-stock company is best managed by a committee chosen by the parties interested in virtue of the fact, and solely in virtue of the fact that it consists in the management, by a few persons, of affairs in which a much greater number are interested along with them. Then, and only then do we have a sound argument by analogy.

To summarise: what we find in Mill is a clear recognition of the difference between Aristotelian analogy and simple similarity. Along with this, illustrations of a type of argument by analogy based on Aristotelian analogy that is completely different from the 'argument by analogy' spelled out by Reid. Most significantly, whereas the other authors we have looked at in this chapter have been largely concerned to warn against the abuses of analogy and the ways in which one can go wrong in the use of argument by analogy, Mill sketches out the conditions under which it is possible to construct a fully valid, and indeed sound, argument by analogy.

CHAPTER 3

Darwin's Analogical Theorising before the Origin

With the ancient background and modern foreground in place, we can ask how Darwin's *Origin* relates to the long- and the medium-run narratives of the last two chapters. This chapter answers with intellectual biography for the years before Darwin wrote the *Origin*. Then three more chapters are on the book itself.

Darwin's Earliest Sustained Causal Four-Term Analogical Reasoning

The intellectual biography needs to be historiographically comprehensive, and so also social and economic. Darwin in February 1835 is weighing possible analogical reasons why species might be, like individuals, intrinsically mortal and dying of old age. He is himself twenty-six years old. After childhood and schooling came five student years, at Edinburgh aiming for a medical career, and at Cambridge with an Anglican priesthood in mind. Then, deferring all career commitments – which he will become too economically and socially privileged to revive – he has done three years as a family-funded gentleman-naturalist on the British Admiralty ship HMS *Beagle*. He was aiding his nation's informal-imperial and import-and-export trading dominance in South America, as Spanish colonial rule has been weakened by Napoleon's recent warring with the mother country. The mission statement for the voyage had identified coral islands among other geological subjects as navigationally pertinent to these national ambitions.[1]

Why then is he thinking about species mortality and extinctions as he is? Most obviously because he has now a commitment, not professional but strongly avocational, to science, explicitly self-identifying as a

[1] For more detailed discussion of Darwin's voyage years theorising, see Hodge (2009a), articles II and III.

geologist. His practical duties include collecting mineral and rock specimens and animal and plant fossils. These duties go far beyond this, to
delineating rock formations and landforms and their submarine extensions. Way beyond these practices, he has become a zealous disciple of the
Scotsman Charles Lyell's views in the three volumes (1830, '32 and '33) of
the *Principles of Geology*. These volumes were all now with Darwin in
the little ship's library of several hundred volumes, mostly on science
and travel.[2]

On theoretical issues, Darwin is often deciding how far he agrees with
Lyell. He will not disagree with Lyell's special creations of immutable
species until mid-1836 and the closing months of the voyage. But on coral
islands he will, by late 1835, have adopted alternatives to Lyell's particular
views, while conforming these alternatives to his mentor's most general
and controversial principles. Those principles concerned the balanced,
untiring, aqueous levelling and igneous unlevelling agencies at work on
and below the earth's surface today, and ever since the laying down of the
oldest-known fossil-bearing rocks of the carboniferous formations. Species
of all the main animal and plant types are originating and going extinct at
all times and places, species of lower types, such as molluscs, lasting longer
than mammal species.[3]

As with coral islands, Darwin's early 1835 alternative to Lyell's views on
species extinctions drew on preoccupations tracing to his extracurricular
Edinburgh zoological mentoring by Robert Grant: preoccupations with
comparing and contrasting artificial and natural, and sexual and asexual
reproduction, limited and unlimited life and individual and associated
(colonial) life in plants and lower animals.[4]

Darwin's February 1835 alternative to Lyell's view of species extinctions
would have a long but limited life. For he gives up this theory and returns
to unqualified acceptance of his geological mentor's view in late September
1838, two years after returning from the voyage. This was, not coincidentally, when he begins two months of changes of mind about species origins
leading in December to his causal four-term selective breeding analogy.[5]

This preliminary sampling of biographical contexts confirms that the
analogical form and content of this extinction theory are decisive for any
understanding of the selective breeding analogy. We can also see why three
historiographical strategies cannot help. One would go on a word search,
for 'analogy', in all the texts from the years before late 1838; but this would
only show that for Darwin, like others, the word had many meanings and

---

[2] Ibid.   [3] Ibid.   [4] Ibid.   [5] Ibid.

uses, most of them not at all aligned with the Aristotelian tradition of reasoning to and from four-term relational analogies. A second would seek some authoritative teaching on the role of analogy in science, read or heard by Darwin, and convincing him that good scientific reasoning conforms to this Aristotelian tradition; but no texts make this credible. Third, perhaps analogical imperatives were in the very air; but even if inhaled this air was too foggy for decisive structural instruction of Darwin's selection analogy.

Back, then, to Darwin's February 1835 geological memo, written when visiting southern South America's west coast, perhaps when sailing between the island of Chiloe – where he had learned about a distinctive apple-tree grafting technique – and Valdivia on the Chilean mainland. For a long time he had been seeking a Lyellian integration of the history of the land and the life on the other, eastern coast. He now reflects that a fossil find made over a year before, putatively of a mastodon, could show that this and other large extinct mammal species had lived on the eastern coastal plains after their elevation from the sea bed, and during the persistence of the extant mollusc species found as fossils embedded in these plains and still living in the ocean. But this conclusion raises an explanatory challenge. For, as Lyell himself would have insisted, the persistence in this area of these mollusc species indicated that no changes had occurred in climate or other local circumstances, changes such as Lyell would have invoked as initiating causes for the mammalian extinctions. So, Darwin turned to the species mortality theory of Giovanni Brocchi, as respectfully rejected by the Italian savant's Scottish friend. 'The following analogy … is a false one' Darwin admitted: but when he considers 'the enormous extension of life of an individual plant seen in grafting of an Apple tree', and 'that all these thousand trees are subject to the duration of life which one bud contained', he sees no great difficulty 'in believing a similar duration might be propagated with true [i.e., sexual] generation'.[6]

An asexual and artificial generation process is causally to the long extending and eventual ending of a domestic tree graft succession at present as a natural, sexual generation is causally to the extending and ending of a wild mammalian species in the past. Four terms and one common repeated causal relation.

Lyell and Brocchi both contributed to the form and content of Darwin's analogy. Although emphatic that Brocchi's theory of extinctions was

---

[6] The February 1835 memo is in CUL MS DAR 42.97-99 and is accessible in the Darwin Online website directed by John van Wyhe. It is analysed and transcribed in Hodge (2009a), article II, pp. 19–22.

unacceptably hypothetical as his own was not, Lyell did concede that Brocchi's alternative could be made more acceptable if certain factual findings were confirmed. If it could be shown that some wild plant species had gradually dwindled and died, 'as sometimes happens to cultivated varieties propagated by cuttings', even though climate, soil and every other circumstance remained the same; and if any animal species 'had perished while the physical condition of the earth, the number and force of its foes, with every other extrinsic cause, remained unaltered', then we would have 'some ground for suspecting that the infirmities of age creep on as naturally on species as upon individuals'.[7]

Darwin's mortality analogy is not like Brocchi's in invoking senescence for species, with declining vital powers preceding extinctions. But he could appeal to precisely those findings that Lyell conceded would support Brocchi's analogy. Even with this support, however, Darwin's analogy is promissory, as he implicitly acknowledges. The propagation of a much extended but limited life is just assumed to be a common consequence of two very different causal processes: generation by grafting and sexual generation. Mill would not think the material circumstance condition has been met; but he would have seen progress in what Darwin does with this analogy two years later and several months after ending his voyage.

In March 1837, Darwin reflected in his *Red Notebook*: 'Propagation whether ordinary hermaphrodite or by cutting an animal in two (gemmiparous by nature or by accident)' always shows us an 'individual divided' either at one time or successively through a long span of years. And he appeals to this generalisation about all generation, in writing that spring on the South American mammal extinctions for the first, 1839, edition of his *Journal of Researches*, and in invoking a notion he knew his grandfather had upheld: that a tree grows as a colony of buds. Occurring with no changes in local conditions, these extinctions, Darwin reflects, 'forcibly' recall – '(I do not wish to draw any close analogy)' – certain fruit trees propagated by grafting and perishing at the same time.[8]

'A fixed and determinate length of life ... has been given to thousands ... of buds' despite being 'produced in long succession'. In a mammal species each individual seems nearly independent of the others, but all may be 'bound together by common laws' as are the myriad buds in

[7] Lyell (1830–1833), vol. II, pp. 128–130; Hodge (2009a), article II, pp. 17–28.
[8] Darwin (1839), pp. 211–212. RN 132. In conformity with standard practice, references to Darwin's notebooks are given in this form as published in Barrett et al. (1987), so that RN 132 is Darwin's MS page 132 in what he called his 'Red Notebook'.

a tree or polyps in a colonial zoophyte. So, extinction comes to such a species as death comes to a tree or to a polyp colony. The new generalisation – that all generation, artificial or natural, simultaneous or successive or whatever, proceeds by divisions of an initial individual – now supports the analogy whereby natural sexual generation is related causally to the extending and ending of a mammalian species, as the art of grafting is causally related to the extended but limited life of a succession of trees propagated from a single initial bud. The common relation is propagation by division of a finite initial source of life, and the common consequence of this common relation is limited duration for the whole divisional succession.[9]

Articulation and Revision of a First Zoonomical System: Summer 1837 to Summer 1838

Darwin does not arrive at his theory of natural selection and its analogy until the winter of late 1838 and early 1839. But any attempt to understand the first formulation of the selection analogy would be frustrated by fast-forwarding from the spring of 1837 and the completion of the species mortality analogy. What comes next may seem to bear only indirectly on the analogy's construction; but this impression should weaken with later hindsight.

Around July 1837, with most of the *Journal of Researches* written, Darwin opened his *Notebook B*, devoted to the laws of life and with the same heading as his grandfather's most controversial book: *Zoonomia*. The grandson's first two-dozen pages outline a comprehensive zoonomical system matching, in overall structure and often in content, Lyell's synopsis of what Lyell called Lamarck's system. This system of theory Lyell rejected unequivocally but respectfully, most generally and explicitly because it falsely held species to be indefinitely mutable over eons, and encouraged misreading the fossil record as traces of an escalation from low life to high; less generally and explicitly, because it implied unacceptable views of humankind.

In articulating his own zoonomical system Darwin was breaking with Lyell more extensively and consequentially than ever before or since, even while remaining loyal to his principles of geological science for the physical world of earthquakes, climate shifts and rock formations. This break in summer 1837 transcended two earlier ones. Most likely in mid-1836, a

---

[9] Darwin (1839), pp. 211–212 and 261–262.

few months before landing back in England, Darwin seems to have thought some biogeographical facts best explained if, contra Lyell, species were mutable so that generic and familial groups of species could have descended from a single ancestral species. In March 1837, in London, in his *Red Notebook*, he had embraced common ancestry for wider groupings and begun integrating this species origins theorising with his species mortality and extinctions theorising.[10]

This integration is greatly amplified in the opening pages of *Notebook B* in the summer. Strikingly, Darwin begins not by comparing, but by following his grandfather in contrasting, sexual with asexual generation. Asexual generation is facsimulative and conservative. Not so sexual generation with its matings, fertilisings and crossing of two parents; and, in the offspring, maturations recapitulating past ancestral progress over eons.

These maturations enable new adaptive, hereditary variations to be acquired in altered circumstances; and, although crossing is counter-innovative when offspring are intermediate in character between the two parents, migration with isolation of a few individuals inbreeding in new circumstances can circumvent this effect of crossing, and so let a new variety form and then diverge enough to be inter-sterile with the parent stock, so becoming no mere variety but a new species. The ramifying reiterations of such species formations make possible the adaptive diversifying descent of a family or class from its common ancestral species; all thanks to those two features, matings and maturations, distinguishing sexual from asexual generations.

So much for adaptive diversification all the way from individual sexual reproductions to interfamilial and wider divergences. What of life's progress as presented in the second movement of Lyell's bipartite synopsis of Lamarck's system? As in that synopsis, progress starts for Darwin with the continual spontaneous generation of monads, the simplest microorganisms, and goes all the way to mammalian perfection. Here Darwin invokes a vast scaling up of his limited vital duration theorising, supposing that the entire progressive generational issue arising from any monadic beginning will have a vast but limited span of life. He draws several corollaries from this limitation, including the conjecture that those lines of progress that have risen most highly must have had quicker changes of species, so explaining why higher species like mammals have shorter species lifetimes than lower species like molluscs.[11]

---

[10] *Notebook* B 1–24.   [11] B 20–22.

Soon after the two-dozen pages of his inaugural zoonomical system of summer 1837, Darwin rejects the vast but limited monad issue lifetime because it falsely entails that all the co-descended species of a genus or family would go extinct simultaneously. He now decides that each species is intrinsically mortal, but that mammal species have shorter lifetimes than mollusc species because wider taxonomic groupings have arisen with more and quicker branchings in the tree of life. This revision and others now made to the second half of his system, on progress, result in a new integration of the two parts. For, with the monads and their peculiar properties no longer invoked, progress from simpler to more complex forms of life becomes a reliable if not invariable consequence of adaptive arboriform diversification, and so of the maturations and crossings distinctive of sexual generations reiterated over eons in constantly and contingently changing environmental circumstances.

Asexual generations extend and so benefit an individual's life; but sexual generating foregoes this individual benefit, for its purpose is to benefit a species by facilitating adaptive changes including those resulting in species descendants. This teleology of sexual generation is integrated with the intrinsic species mortality theorising in summer 1837. An asexual succession of grafted apple trees can be saved from death by sexual crossing which can extend life for further generations. So, the sexual succession of a higher animal species can end in childless extinction thanks to some very adverse change in conditions before its lifespan runs out, or, again, with conditions constant, when that span is eventually spent; while in less adversely changing conditions, Darwin now emphasises, a species can adapt and give rise to one or more offspring species each with its own new lease of life. Later in 1837 he extends this likening of species being born, living and dying to individuals doing so. Two human families may differ in the number of descendants they have at some future time thanks to good or bad luck with hereditary diseases and other such contingencies. Likewise for lucky winners and unlucky losers among ancestral species and their ramifying species descendants.[12]

Over the next two decades before the *Origin*, Darwin will make almost no radical revisions to this theorising about these arboriform patterns and processes. The two exceptions are rejection of intrinsic species mortality and injection of natural selection, both revisions coming in the last quarter of the following year, 1838.

[12] Hodge (2013a), pp. 66–70.

In early 1838, as he begins *Notebook B*'s successor, *Notebook C*, he sees himself as knowledgeable, not ignorant, concerning the causes of the origins and extinctions of species, their quasi-individual births and deaths, and their adaptive successes and failures, for he has related these events and processes to an integrated understanding of all kinds of generation in animals and plants. He was also now even more aware that his own intellectual ancestors included not only Grant and Lyell but his grandfather, whose *Zoonomia*'s most notorious chapter was headed 'generation'; and Lamarck whose theorising, like the *Zoonomia*, had been endorsed by the zoologically heretical Grant in discussions with the student Darwin in Edinburgh. These biographical generalisations about Darwin's bodily practices, in his covert notebook brainwork, confirm that new notions arising in the months of late 1838 and early 1839 are not always filling empty cognitive gaps, but are promising to amplify theorising already in play.

Throughout the two years of *Notebooks B–E*, Darwin was often comparing variety formation in domestic and in wild species; and so following Lyell who had drawn on decades of researches comparing the racial varieties of man with dog and livestock breeds; researches prominent in Buffon and Blumenbach and continuing with Lawrence and Prichard; researches Lyell held to confirm, contra Lamarck and others, that intraspecific varietal diversification never leads to interspecific diversification of humans or of animals or plants domestic or wild.

From autumn 1837 on, Darwin's species formation theorising is often directed to explaining two permanent changes: adaptive divergence of structures and instincts, and loss of fertility in crossing. Cases of nonblending of parental characters, especially in human interracial crosses, indicate incipient constitutional incompatibility. Ornithologist William Yarrell told Darwin that if two breeds of domestic animals are crossed the offspring have the characters of the more ancient breed. Elaborating many corollaries from this law, Darwin concluded that over many generations any hereditary characters must become so firmly and powerfully embedded in the constitution that a blending compromise between two very old breeds is impossible. This conclusion gave him a new way of comparing and contrasting species formation in the wild and variety formation under domestication. Some domestic breeds, although markedly different in bodies and habits, interbreed readily whereas wild species differing that much do not. Darwin took it that domestication itself, this unnatural condition, vitiated the instinctive aversion to interbreeding shown by even very similar wild species. Con-specific domestic breeds support his theory of wild species formation by showing that new, slightly divergent wild varieties would avoid interbreeding, and so could go on to become incapable of interbreeding and become species.

From the early months of 1838, Darwin persistently contrasted two sorts of domestic breeds: natural varieties and artificial ones. Natural varieties of domestic sheep and cattle, especially, can be due to natural causes, not human artifice, and are often local varieties, regionally isolated and not interbreeding with others and diverging as they adapt slowly over many generations to local conditions of soil, climate and so on. By contrast artificial varieties are not adaptive and may even be monstrous. They are distinguished by variations arising as rare maturational accidents, variations that only persist thanks to the human art of picking, as Darwin calls selective breeding. Often made in a few generations, these varieties could never be formed and flourish in the wild without benefit of this human art. Darwin's knowledge of selective breeding and convictions about its efficacy were consolidated in the summer of 1838 by reading an authoritative pamphlet by Lord John Sebright, and this knowledge will endure for all the years to come. Sebright's own epitome could be Darwin's from now on: 'the art of breeding ... consists in the selection of males and females, intended to breed together, in reference to each other's merits and defects'. However, as Darwin reads about selective breeding, or picking, at this time he is confirmed in his conviction that there is nothing like picking at work in the forming of species in the wild, and that these formations are to be compared with natural variety formation in domestic species and contrasted with the making of artificial varieties.[13]

Wild species formation is an adaptive achievement made possible by adaptive and not monstrous variations. If a puppy moving to a cold climate grows thicker fur than its parents, this is an adaptive variation induced by these conditions and advantageous. Thicker fur on a puppy born in a warmer country is a monstrous variation, a response, even an adaptive response, to rare, unhealthy uterine conditions. All these variations are made possible by sexual generation, but only the adaptive ones contribute to species formations; rare monstrous ones are blended out in crossing and are less able to survive and reproduce themselves.

Darwin thinks adaptive variations and so species formations are often initiated by changes in habits and so in the use of organs. If all the jaguars in a region swim for fish prey on their country becoming flooded, then a

[13] C 133.

new variety with webbed feet could arise through the inheritance of this acquired character.

In his *Notebook D*, filled from July to October 1838, there are no revisions concerning this generational and ecological view of variety and species formation in the wild. This notebook is dominated by theorising about sexual generation including its providential teleology. In ontogeny and phylogeny he holds that hermaphroditic sexuality precedes the separation of the sexes found in higher animals; and that any unfertilised egg in a female is like an asexual bud and so incapable of acquiring novel hereditary characters due to pre- or post-natal influences, these acquisitions being the very purpose of sexual generations and needed for species changes, adaptations and progress. Here, Darwin appeals to asymmetries between parental powers, with novel hereditary characters being mostly acquired and impressed on offspring by males. Now, with these generational and ecological themes briefly surveyed, we can move to the months of theorising that lead Darwin to his most sustained and familiar analogy.

On the Origin of the Selection Theory with No Selection Analogy

The topics of the next two sections have become historiographically hazardous. Here are the warnings we have given ourselves. First, Darwin's many autobiographical recollections are guileless but often misleading, including a note in his *Notebook D*, implying that late in September 1838, he 'first thought of selection owing to struggle'.[14] Second, as with other notebooks, so with *D* and *E*, some excised pages are missing; the story would likely look different if we had them. Third, dating Darwin's notebook entries and marginalia is not always possible; narratives grounded in what is securely dateable take precedence. Fourth, the famous sentences prompted by Darwin's late September reading of Robert Malthus on population are, except for a closing one, not about species origins but extinctions. Fifth, this sentence, inserted interlinearly, but most probably at that time, shows Darwin responding not to Malthus's moralistic pessimism about English industrial and urban life, but to his upbeat providential teleology and theodicy for ancient empires and, by implication, for modern colonial invasions and settlements. Sixth, Darwin arrived at his selection analogy no earlier than late November; and there is no prior reversal of his longstanding comparing of wild species

---

[14] *Notebook D*: inside front cover.

formations with the formation of natural varieties of domestic species, and his contrasting both of these natural formations with the making by selection of artificial varieties of domestic species. Seventh, Darwin did not arrive at his analogy by reasoning thus: artificial selection is known to produce domestic varieties; wild species are like such varieties; like effects have like causes, so, there exists in the wild a similar process of natural selection that can produce species and has done so. Eighth, Darwin's personal consultations with animal breeding experts contributed little to his initial arrival at his analogy. Ninth, a word search for natural selection does not help. At least one author, unknown to Darwin, had scooped him in talking metaphorically of a natural process of selection. Darwin probably first used the term 'natural selection' in a manuscript of 1841, long after deploying the selection analogy using other wording from late 1838 on. The presence of sorting, picking and sifting talk is what rewards attention, not the absence of selecting talk.[15]

Finally, Darwin's arrival at his analogy shows him integrating various comparisons and contrasts about the causes and effects of various causal relations. Malthusian excess fertility causes the struggle for existence in the wild, which causes natural selection, which causes adaptive species diversifications – just as stockbreeders cause artificial selection which causes varietal diversifications adapted to human uses and fancies. The struggle is causally related to animals in the woods as stockbreeders are causally related to animals on farms. Here, the two selective processes are not related items, not relata, but relations, for these two similar causal processes are relating the utterly dissimilar struggle and stockbreeder to their respective effects, the forming of wild and domestic varieties. Although having wholly unalike causes, the two relations, natural and artificial selection, are themselves sufficiently alike to have similar causal consequences. Our hazard warnings can end with a resolution to study what was and wasn't new to Darwin in these causal–relational reasonings, while avoiding anachronistic exegeses.

On 28 September 1838, engaging Malthus's insistence that fertility far exceeds replacing losses from deaths in any human or animal population, Darwin argues that this excess fertility makes competing species liable to extinctions initiated by even very slight changes in conditions. Citing exceptional human populations doubling in twenty-five years, he is alerted to food limitations and other checks normally preventing such potential increases becoming actual. There is a force, he says, 'like a hundred

[15] Hodge (1992a), pp. 213 and 215.

thousand wedges' tending to force 'every kind of adapted structure into gaps in the oeconomy of nature, or rather forming gaps by forcing out weaker ones'. In this simile, no log is split and the wedges are not hammered but self-driving; and the width of the wedge where it enters the log represents actual population numbers, and the greater width outside the higher potential numbers. The inward driving of winning wedges requires losers to be forced outward; so, most species, Darwin concludes, are pressed hard in fragile competitive balances that minor environmental shifts can upset, causing total population losses for losing species. Comparing the *Journal of Researches* of 1845 with the first, 1839, edition, shows this reflection returning him to Lyell's view of extinctions, and away from his 1835 alternative view that some extinctions are due to intrinsic species mortality. This generational theory and its mortality analogy are now replaced by Lyell's vindicated ecological theorising.[16]

This much for the losing species, but what of the winners? In the interlinear sentence Darwin asserts that the 'final cause', the divinely intended good effect of all this wedging, is to 'sort out', to retain, 'fitting structure' and so 'adapt' structure to these changes in conditions. Structure is thereby adaptively enhanced in animals, just as, he notes, Malthus has shown how the 'energy' of victorious ancient peoples was providentially enhanced by life and death struggles, when excessive fertility and consequent food shortages forced their tribal migrations and imperial invasions onto contested, occupied ground. Here, Darwin responds to Malthus as one theist extending another's teleology and theodicy for reproductive and imperial drives and powers. The wedging simile itself occurs more briefly in later texts including the *Origin*, but is never elaborated as a metaphor or developed into any analogically structured reasoning. The sorting talk about the wedging's good effect would eventually have a far more fertile future.[17]

This Malthusian sorting goes on between and within species, Darwin seems to imply; but he makes here no comparison with the picking or selecting practised by stock breeders. Nor is there any rethinking on how sexual generation ensures adaptive change in altered conditions. He emphasises over the next two months what this sorting entails for advantageous variations acquired in individual maturations, and for exchanges of new species for old in changing conditions over long ages of time. He concludes that only a structural variation advantageous for the whole lifetime of an individual will be retained in the Malthusian crush of

[16] D 134–135 and Darwin (1845), pp. 173–176.   [17] D 135.

population over many generations; variations adaptive to foetal circumstances alone will not be; and retained variations, eventually becoming strongly hereditary, can be accumulated in prolonged changes. Thus do his new Malthusian insights fit with his earlier views on both adaptation and progress.

In late November, Tuesday the 27th very probably, in his *Notebook N*, (sequel to *M* on metaphysics, on mind, that is, including morality) Darwin relates long-run adaptation and progress to his first explicit contrast between two principles that explain changes in structures in the short run. One principle is familiar enough: an adult father blacksmith, thanks to the inherited effects of his habits has children (well, sons anyway) with strong arms. The other 'principle' has no precise precedent: any children whom chance has produced with strong arms outlive weaker kids. The contrast is direct because chance production means, as it has all along for Darwin, production by small, rare, hidden causes effective prenatally, so that the opposite of chance is postnatal habits. New here is the conviction that those products of chance having the same benefits as the effects of habits can contribute to adaptive change because, although rare, individuals with such beneficial variant structures will survive over future generations at others' expense. However, Darwin acknowledges a difficulty in deciding which adaptive structures – and instincts too, because these two principles apply, he notes, to brain and so mind changes – have been due to which of the two principles.[18]

A few days later, by the Sunday after that Tuesday, he is, in *Notebook E*, again considering principles. This time there are three and they can, he says account for everything. He may have wanted these three principles to subsume the earlier pair, so circumventing the difficulty of deciding which changes to ascribe to which one of that pair. Strikingly none of the three principles is new to him: that grandchildren resemble grandfathers; that there is variation in changing circumstances, and that fertility entails potential population numbers greater than limiting resources, especially food, can support. Darwin was at this time reading books by Herschel and Whewell that may have stoked his longstanding Newton envy with an ambition to frame a few maximally explanatory principles.[19]

This triumphant claim for these three principles of heredity, variation and the struggle for existence can be fairly deemed the moment when Darwin first embraces natural selection but with one caveat: this is still a theory of Malthusian sorting rather than a theory of 'natural selection', for

[18] N 42.   [19] E 58.

no selection analogy has had a role in Darwin's arrival at this theory. Nor is there any implicit selection analogy argumentation. But there will be, perhaps within a few days and no more than a fortnight.

What, then, is most probably going to be Darwin's path to his analogy? We have confirmed that Darwin did not reach it by reasoning that domestic varieties are made by selective breeding; that wild varieties are like domestic varieties; and that like effects have like causes; so wild varieties are made by a natural selection like artificial selection. Had he done so, he would have been inferring the existence of selection in the wild from its existence on the farm. But he wasn't. From the first Malthusian reflections he reasoned that, owing to the struggle for survival in the wild, sorting exists in the wild and causes adaptive change.

He had earlier learned that owing to stockbreeding practices there is selective breeding on farms causing diversification of domestic varieties. We see next that only after two months of further sporadic Malthusian musings does he come to see the discriminating sorting effects of the wild struggle as resembling the selective effects of the breeders' art. So, contrary to what he'd long thought, there is a process like this selection by man going on in the wild. This new acceptance that selection exists in the wild is not an inference from the existence of selection on the farm, but is inferred from the consequences of the existence of the struggle for survival in the wild. Moreover, this conclusion requires a reversal of the long-standing comparison of wild species with natural rather than artificial varieties of domestic varieties because, now, the struggle, in its selective actions and their consequences, is seen to be causally to wild animals as the stockbreeder is causally to animals on the farm. Same causal relation and same consequences. This is the four-term causal–relational proportional comparison constituting the selection analogy.

Adding an Analogy without Subtracting Any Theory

We have to ask what first prompted Darwin to make this relational comparison of the wild, natural sorting and the artificial, domestic selection or picking. This is a question about two legacies from two Roberts, population theorist Malthus and sheep-breeder practitioner Bakewell. The documentation only allows conjecture, but does make one line of guessing instructive.

The new comparison led Darwin to view positively both the natural sorting and the domestic picking, so it included a move away from the negative take on picking as a means whereby monstrous and unadaptive varieties could be made by man. This move could have been encouraged by Darwin's privileged rural preoccupation at this time with the hereditary instincts of sporting breeds of dogs, especially greyhounds, which, in being fitted by man for prize-winning excellence in tracking and catching hares, have been made more, not less, adapted for life in the wild. His earliest surviving allusion to picking by nature as well as man, on or soon after 4 December, offers another example: dogs bred and trained to retrieve shot birds from lakes and rivers. 'Are the feet of water-dogs . . . more webbed' than in other breeds, he asks? Adding that 'if nature had had the picking she would make such a variety . . . far more easily than man'. The echoes of those jaguars adapting to fishing are telling. And the contrasting of the superior power of nature's over man's picking is explicit.[20] As a friend has quipped: having travelled the world on the *Beagle*, Darwin may have gone on – ever doggedly – to natural selection by greyhound.

By the middle of December the comparing and contrasting of the two pickings is well on. Though less powerful, man's picking is more knowable and so can be a domestic, artificial model for the new analogy's wild, natural target. It's 'a beautiful part of my theory' that domesticated races are made by the very 'same means as species', although species are made 'far more perfectly & infinitely slower'. No domesticated animal is perfectly adapted to external conditions. Only nature's prolonged picking can cause 'those ancient and perfectly adapted races' that count as species. In his analogical comparisons of Bakewellian selection and Malthusian sorting over the next three months to mid-March, Darwin was often developing new insights concerning variations that are rare, slight and 'accidental', due to 'chance', but are nonetheless able to contribute to adaptive species formations thanks to sorting or sifting in the wild being like artful domestic picking by man.[21]

As soon as Darwin has his new analogy in late November or early December, he starts learning about natural selection by analogical reasonings from what he is learning about artificial selection. The two learnings are integrated because, as we will see him explicitly reflecting in March in rehearsing potentially public argumentation, unlike nature's selection man's selection can be learned about from the experiences of authoritative practitioners, being observationally accessible as natural selection is not. The analogy makes possible learning from observable art about what is inferably natural.

[20] E 63. For a more detailed account, see Hodge (2009a), article V, pp. 200–202.
[21] E 72. Hodge (2009a), article V, pp. 202–203.

One expectation is not fulfilled. Knowing that Darwin put questions to animal breeders, biographers expect to find him learning about selective breeding from this exercise at this time. But the evidence shows otherwise. Most probably in 1838, and before arriving at the analogy, he had drafted questions to discuss with an animal breeder and friend of his father, Rice Wynne; and those questions are predictably enough mostly about Yarrell's law and the corollaries Darwin had developed. Less predictably, the printed questionnaire sent to various breeders early in 1839 is also dominated by those same concerns together with longstanding concerns about inbreeding, anxieties indeed, now that Darwin is married to his cousin. We learn very little from these inquiries about Darwin's earliest analogically structured and instructed learning about artificial and natural selection.[22]

Going back then to *Notebook E*, and also to Darwin's revealing notes on a theological treatise by John MacCulloch, most probably from the first four months of the new year, we see that learning about the relationship between the accidental and the adaptive is continuing to preoccupy this private, covert theorist after his (and Abraham Lincoln's) thirtieth birthday on 12 February; and we see that a leading theme now invokes the ancient threefold distinction between the accidental, the necessary and the adaptive.[23]

The perfect adaptation of species is due to nature's selection being supremely discriminating and comprehensive in its consequences for the whole body, inside and out, and whole life before and after birth. Outward greyhound form might be made by man's selection away from all hares and hunting, but a perfected race would be formed in nature only through the perfecting selection that living by hunting would entail. So, the selection analogy itself implied that the adaptive perfection of species may not be due to any difference between variation in the wild and under domestication, for species may be formed by chance or accidental variations just as domestic breeds often are. Darwin now favours the contribution of chance variation to adaptation, over the other blacksmith principle, especially in extending these analogical arguments to cases such as seed dispersal structures in plants where he saw no habits having inherited effects, nor any plausible way for them to arise in necessary adaptations as with the puppy growing thicker fur in a colder clime. When reading Darwin pursuing such themes at this time, we may think that much of what he will say about artificial and natural selection over the coming decades traces to these

---

[22] Hodge (2009a), article V, pp. 197–203.   [23] Ibid.

months. If we refrain from misleading anachronisms, this is not a temptation to resist.[24]

Malthus had prompted the notebook Darwin to take seriously small causal differences having large accumulated consequences. Two regions may differ only very little in their soil and climate and so on, but Malthusian maths shows why a species is more populous in one region, because small differences in survival rate have very different compounded outcomes over even a few generations, just as, diachronically, minor changes in conditions can cause small but consequential changes in those rates. Conversely, within a species very small differences in survival rates between two varieties can lead to increasingly divergent future frequencies. As Darwin often reflects, in the lives and deaths of reproductive creatures, especially though not only in the wild, a grain of imbalance can be very consequential.

In his notes on John MacCulloch's theological treatise, he sees his 'theory of grain of small advantages' explaining the curling seed pod valves of broom plants, on the presumption that these are not 'necessary adaptations' but 'accidental', for 'they would not be detrimental accidents and domesticated variations show us accidents may become hereditary ... if man takes care they are not detrimental'. In his notebook he acknowledges the difficulty in believing 'in the dreadful but quiet war of organic beings' when he goes into the 'peaceful woods and smiling fields'. The difficulty is overcome by considering the difference between the very restricted actual range of a species and the range it would have if no other competing species were limiting its spread. There is, he says, 'a contest and a grain of sand turns the balance'. Adaptive species formations will be slow and gradual because advantageous variants are often only slightly so, and would arise initially by chance in only a few individuals. Dogs with hereditarily longer legs might take ten thousand years to get the upper hand in the Malthusian rush for life.[25]

There are then appeals to chance and to chances to be explicitly distinguished. Variations, in legs, say, may arise by chance, by accident, but, if, among them, there are increases in leg length causally conferring greater chances of survival and reproduction, those greater chances and their consequences over successive generations will not be accidental. Around mid-March, in more notes on MacCulloch, Darwin is looking forward to public argumentation on behalf of his theory and its analogical support. He resolves to get 'instances of adaptation in domestic varieties

---

[24] Ibid.   [25] Ibid.

such as 'greyhound to hare ... waterdog hair to water' and several other
plant and animal examples that are just as much adaptations as the
woodpecker's body and habits are, for 'here we see means' of adaptive
formation as we do not with the wild bird species. Even more explicitly, in
Notebook E, he reflects at this time on how his introduction of his theory
will begin by appealing to the selection analogy. He will start by saying
once again that domestic varieties are made in two ways. One is when an
entire portion of a species is subject to the same influence of conditions as
happens on its moving from one country to another. But he resolves to
insist: the greyhound, race-horse and pouter pigeon 'have not been
thus produced', rather 'by training, and crossing and keeping breed
pure'. And in plants, likewise, *effectually* the offspring are picked and
not allowed to cross'. The decisive question is, therefore, 'has nature
any process analogous – if so she can produce great ends', and 'but
how – even if placed on Isld. If etc etc ... Here give my theory –
excellently true theory.'[26]

This memorandum epitomises the argument strategy later followed by
Darwin in the opening sections of his *Sketch* of 1842; for he starts in 1842,
as he did in *Notebook B* in summer 1837, with individuals varying
heritably and adaptively in new conditions thanks to sexual as contrasted
with asexual generation, before going on to the blending out of this
variation due to crossing. Here, in 1837, Darwin introduced isolation
and consequent inbreeding counteracting the conservative action of
crossing. In 1842 he introduces a series of 'ifs' matching those promised
in March 1839, in considering how human selection could counteract
these counter-innovative effects. The conclusion from all the 'ifs' is that
even if human selection, aided by isolation, operated to the full extent of
its power, it would not produce very much permanent adaptive change,
because man's judgement is poor, is restricted to external characters and
cannot lead to a race fitted to all the conditions of its life. Thus is set the
stage for natural selection with the superior powers needed to achieve the
results man's selection cannot.[27]

At this moment, both the 1842 *Sketch* and 1844 *Essay* have a theological
excursus after artificial selection is introduced and before natural selection
is. These passages begin by imagining the selective breeding powers of an
imaginary being (with capital B in 1844), infinitely smarter than any
humans, acting super-intelligently and super-discriminatingly on the lives
and deaths of animals and plants, and so with imagined powers vastly

[26] E 118.   [27] Darwin (1842), pp. 45–46; (1844), pp. 114–115.

superior to those of any human breeder. Thus far, Darwin's Christian God, who is not imaginary and always has a capital initial, has not been considered. But Darwin, still the convinced theist he will be for years to come, does now get real and theological, for he asks if any such superior selection actually exists and, invoking the venerable distinction between direct unmediated and indirect mediated divine action, he says the real Creator (capital C in both texts) may be assumed to work, as usual, indirectly by natural, secondary intermediate means. The question is then whether there exist any selective breeding means mediating between God and his creatures. The reader easily guesses what's coming next: natural selection with its vast superiority over human selection. And sure enough there follows the section on natural selection, opening with De Candolle on the war of nature and Malthus on the war's causes.

Darwin has here distinguished two selective breedings to compare and contrast with human selection: the imagined selection of the imaginary Being and the real natural selection mediating between the real Creator God and his creatures. As selective breedings both are comparable in kind to artificial selection by man; and comparable with each other in far exceeding in degree the powers of mere human selection. But, there is a decisive contrast: the imaginary Being is explicitly imagined to select, as men do, by acts of foresight and will; while the struggle for existence and its Malthusian causes, do not. This theological excursus supports the ensuing emphasis on the enduring, ubiquitous presence and great adaptive, diversifying powers of natural selection, while crediting mentality only metaphorically to nature, as distinct from nature's essentially mindful author, God. Millenia of orthodox Christian opposition to pantheistic heresies are to be reaffirmed, with no need to rethink the selection analogy.

From the 1842 *Sketch* and Beyond

Notebook E ends in the summer of 1839; and we have seen that three years later Darwin writes his first draft of what has been known, since its posthumous publication in 1909, as his *Sketch* of 1842, identifiable in hindsight as the earliest known textual ancestor of the *Origin*. Documentation of his thinking about natural selection during those three years is scant; and the *Sketch* text is messy and scrappy. But comparing *Notebook E* with three distinct *Sketch* drafts, from 1842 and the two following years, confirms that the theory and its analogy are fundamentally the same in 1842 as they were within a few weeks of their first formulations; and the composition of these draft texts is conformed to the same compositional ideal that the *Origin* will be conformed to in 1859.[28]

This vera causa (true cause) compositional ideal complements the vera causa evidential ideal. In accord with both Darwin explicitly plans to divide his argumentation – on behalf of common descent by means of natural selection, branching natural selection – into two halves. In the first half the evidential case is made for the existence of branching natural selection and for its power, its adequacy to produce and adaptively diversify new species from old. In the second, the explanatory virtue of this theory is demonstrated by showing how many different kinds of facts, especially paleontological, biogeographical, embryological and morphological facts, it can explain, and so why branching natural selection is very probably responsible for the production of species, families and so on in the earth's long past. In sum, existence, adequacy, responsibility. It is, it can and so could have, and it did.[29]

This compositional ideal complements the evidential ideal because the vera causa evidential ideal held that a causal–explanatory theory should have the existence of its cause evidenced independently of those facts it is being used to explain, facts that are thereby evidencing its responsibility. And the natural ordering of the three evidential cases, from existence through adequacy to responsibility is supported by precedents in Newton and Lyell especially, and by the reflection that only a real, existing cause can be adequate and only an existing and adequate cause can be responsible.[30]

Darwin argued for branching natural selection in this way thanks remotely to Thomas Reid who wrote early in his 1785 book of *Essays on the Intellectual Powers of Man* a short chapter on hypotheses. Being a strict empiricist he was against granting them any serious credence. But in commenting on one of Newton's methodological dicta he specified two conditions to be met if a causal explanation is at all credible: there must be sufficient evidence that the cause invoked does really exist, and the effect to be explained must necessarily follow from it.[31]

What is distinctive here is the requirement of independent evidence for the existence of the cause. A cause meeting this requirement will be deemed a true, real, known and existing cause. It had long been a commonplace that evidence of causal adequacy was required even in a purely conjectural hypothesis. And it was long taught that a good hypothesis should display explanatory virtue in accounting for many different

[28] Hodge (2008), article VIII, pp. 237–246.   [29] Ibid.; Kohn (1982).   [30] Ibid.   [31] Ibid.

kinds of facts. This trio of conditions constitutes the complete Reidian evidential ideal.

Darwin's first encounter with this Reidian legacy came most likely before leaving on his voyage when he read enthusiastically John Herschel's 1830 *Preliminary Discourse on the Study of Natural Philosophy.* He was probably impressed by Herschel's defence of Lyell's geology as conforming to the vera causa ideal; and this favourable impression was strengthened in the voyage years as Darwin aligned himself with Lyell's *Principles.* The consensus between Herschel and Lyell and Darwin was apparently alluded to in Darwin's discussions with Herschel in South Africa in mid-1836. It is implicitly prominent again in Darwin's writing on erratic rocks in autumn 1838. Darwin was then rereading both Herschel and Whewell, and surely noticing that on the vera causa ideal generally, and on Lyell's appeals to it as foundational for his geology, Whewell was explicitly negative. Whewell's alternative evidential ideal, only published in full in 1840, was his consilience of inductions doctrine. Darwin never joined Whewell in disagreeing with Herschel and Lyell; and so never accepted this Whewellian consilience doctrine as an alternative to their Reidian vera causa ideal.[32]

There is a nice complication in Darwin's relation to Reid's *Essays.* Within a page of specifying his vera causa desiderata, Reid starts a chapter on analogy, where he quickly comes to his leading theses on analogical similitude as exemplified by an argument for life on other planets. Here is one more compelling reason for seeing Darwin's selection analogy not as Reidian but as Aristotelian: Reid himself makes no connection between vera causa evidencing and analogical argument. For Herschel, analogical argument can make probable the existence of some cause. Consider, he says, the tension felt when we whirl a stone on a string; and consider also that similar effects have similar causes. Together this fact and this principle make probable the existence of a centrally directed force keeping the moon in its orbit around the earth, a force that is to the moon as the string tension is to the stone.

For Darwin, the existence of natural selection is not argued for as follows: domestic varieties are formed by selective breeding; similar effects have similar causes; wild varieties are like domestic ones, therefore there exists some cause in the wild like man's selective breeding, a natural selective breeding which makes wild varieties. No, he reasoned thus: Owing to Malthusian excess fertility there exists a struggle for existence

[32] Hodge (2008), articles VIII, IX, X and XI.

in nature, and its effects are a natural selective breeding similar in its effects to artificial selection. So, owing to the wild struggle for existence there exists a natural cause with very different causes from the causes of artificial selection but similar in its effects. The struggle is then causally to wild varieties as man's selection is to domestic ones, but although similar in kind to the effects of artificial selection, the power and so effects of natural selection vastly exceed those of artificial selection.

The selection analogy is not evidencing the existence of natural selection, so it is not showing that it is a vera causa, a really existing cause (the argument from the effects of the struggle takes care of that). It is showing not that natural selection is a really existing vera causa, but that natural selection is an adequate, competent cause of adaptive species formation and diversification. The selection analogy shows that natural selection could have been responsible for the formation of extinct and extant species; and, furthermore, that many diverse facts about species, facts from biogeography, from morphology and so on are explicable as resulting from branching natural selection, showing that natural selection was probably actually responsible.

At no point was Darwin integrating vera causa evidential ideals and causal proportionality–analogy evidential ideals in ways that he could – much less could only – have been learned from Herschel. And indeed there are no credible documentary traces of such influences and debts.[33]

Darwin's integration of his Aristotelian selection analogy with the vera causa ideal proceeds in the *Sketch* of 1842, as it will in the sequels including the *Origin*, through his twofold partition. First he assembles and evidences his explanatory resources. Then, second, he defends and deploys them. A couple of canonical works had this same structure: Newton's *Principia Mathematica* and Lyell's *Principles of Geology*; and Lyell gave his epitome of Lamarck's theorising (which he rejected) a similar two-part structuring that was not Lamarck's own. Lyell's dozen or so pages on Lamarck have an opening part, on the causes working at present to adaptively diversify species, and a closing part on how progressive change has gone over vast past eons; all in accord with the vera causa compositional ideal which, in Lyellian geology, required first examining present causes and, then referring past effects to those present causes. We have seen how Darwin opened his *Notebook B* with a modified version of this two-part system.[34]

A year later, in summer 1838, and half a year before arriving at natural selection, Darwin is again explicitly planning a bipartite exposition of his

---

[33] Ibid. The above corrects some misleading suggestions that may have been given by these articles.
[34] B 1–24.

conclusions. He knows that ideally a causal theory offered to explain certain kinds of facts should be supported evidentially in two ways: independently of those facts and by showing how well it explains them; and so he resolves to argue for his branching species propagation theory appropriately: first, by evidencing the peculiar powers of sexual generation, including Yarrellian hereditary constitutional embedding and the adaptive diversification of domesticated species into natural varieties; then, second, by showing how this theory can explain, can connect and make intelligible, many different kinds of facts about species: biogeographical, paleontological, comparative embryological facts especially.[35]

Darwin was therefore committed to both the compositional and the evidential vera causa ideals long before reaching his selection analogy; and the new analogy contributed to his conforming to those ideals right away and ever thereafter. His expositions in 1842 and 1844 start with hereditary variation in domestic animals and plants. This variation, more abundant on the farm than in the woods, is probably due to changed conditions of life, especially nutritional changes. Without selective breeding by man it yields little permanent adaptive diversification, but with selective breeding it is quite otherwise. Shift now to variation in the wild. This is less abundant and less easily evidenced from direct observation. But geology, Lyell's that is, shows that species are always subject to fluctuating conditions; and a straightforward causal generalisation from knowledge of variation under domestication can confirm that hereditary variation will result from those fluctuations. Thus far no invocation of any selection analogy. Nor with the next step. Without selection this hereditary variation in the wild yields little change. But there is a struggle for existence and therefore a natural selective breeding as its consequence. So, now finally comes the analogy, because this struggle is shown not merely to exist, but to be causally related to wild animals as the human breeder is to his stock. That is the principal argument to and for the selection analogy; and the principal argument from the analogy shows that this selection has powers and causal adequacy of the same kind but much greater in degree than man's has.

Darwin's Analogy before the *Origin*

So far so familiar. With this argumentation to and from the analogy, the evidencing of natural selection's existence and adequacy is complete, and likewise then the assembling of the causal–explanatory resources to be

---

[35] Hodge (2013a).

deployed in the second part of the 1842 composition. The composing of the 1859 *Origin* is still a decade and a half away. But the selection analogy and the principal arguments to it and from it are not going to change in form or content. Darwin's life as theorist was active enough in those years. But none of the three new insights that might be thought to prompt such revisions does in fact do so. Consider in turn his formulation of his theory of generation, pangenesis, dating most likely to about 1841 but only published in 1868; his elaboration of his sexual selection theorising as completed around 1856–1858; and his formulation, about the same time, of his principle of divergence.

From 1835 to 1837 Darwin had been comparing sexual and asexual generation. Then for two years he retained the comparisons but drew even more fundamental contrasts. By 1841 he was retaining the contrasts but insisting again that all generation was ultimately alike; and this principal thesis of his hypothesis of pangenesis was still paramount in 1868. He never explicitly integrated pangenesis with natural selection. The distinctive powers of sexual, as contrasted with asexual, modes of generation were always invoked by Darwin, before and after that hypothesis's genesis, as causally necessary for both natural and artificial selection, because uniquely productive of hereditary variation. But the theory of natural selection was not a theory of generation. Pangenesis was, just as Darwin's 1837 generational birth, life and analogical mortality of species theorising had been. Natural selection, with its Malthusian analogy, complemented Lyell's theory of species extinctions in being ecological, not generational. Ecological natural selection and generational pangenesis could have been fitted together, but their common authorial parent apparently saw no benefit to either in so doing.[36]

If pangenesis was too remote from the selection analogy to require its revision, sexual selection was too close. Sexual selection was about struggles for mates, males or females competing with other individuals of the same sex; in male combats, with horns, tusks or spurs fitting this end, or in their singing or displaying to females whose choices decide which males are winners. In 1844 Darwin compared such natural struggles among males to the effects wrought by agriculturalists attending less to the many young domestic animals they breed, and more to the occasional use of a choice male. By 1859, he could add that if a man's persistent selection can quickly make his bantams handsome, then wild female birds can work such effects by consistently selecting the most beautiful males according to these

---

[36] Hodge (2009a), article VI, pp. 227–243.

females' standard of beauty. In all these sexual selections the competitive winning of mates is to animals in the wild as man's artful selection is to domestic livestock; and so this analogy is structured and functions like the natural and artificial selection analogy, with two qualifications: sexual selection – picture a peacock's tale – can sometimes explain what natural selection cannot; and sexual selections in the wild may be less rather than more powerful than sexual selections on the farm. Sexual selections in the wild are also less severe and less powerful than natural selections; losers in male combats or female choices may miss out on some matings, whereas the struggles causing natural selection can cost lives.[37]

The *Sketch* and the *Essay* follow the early 1839 notebook theorising in concentrating on the character of those races formed by natural selection. They are, Darwin says, slowly formed and perfectly adapted races distinguished by sufficient hereditary and adaptive differences to count no longer as mere varieties, but as varieties that have become species.

In the 1850s he asks whether natural selection will turn varieties into species by increasing their differences from one another. His positive answer invokes an ecological–agronomical version of an economic generalisation: the productive advantages of a division of labour, ensuring that many different species will yield more crop from any patch of land than only a few will. As a corollary, Darwin concludes that specialisation is advantageous, and that natural selection will favour adaptive extremes just as artificial selection often has. Horse breeders, wanting draught horses stronger and race horses swifter than intermediate breeds, have favoured and bred consistently and persistently for extremes. Specialisation is itself enhanced by structural and functional differentiation; animals with some appendages fitted for feeding, others for running, are favoured over those whose appendages are all structured and functioning alike. Thanks to the Estonian von Baer and his French follower Milne-Edwards, this take on differentiation led Darwin to favour embryonic characters and their homologies in classification, including barnacle taxonomy. And it led him too to favour differentiation as a criterion of ontogenetic and phylogenetic progress. Natural selection causes phylogenetic progress in causing divergent adaptations. Distinguishing the types of differentiation from degrees of differentiation allowed progress to accompany branching. A single ancestral species could have many descendants equally higher in their degree of differentiation than this ancestor, but differing in how they are adaptively diversified and so structurally and functionally

---

[37] Darwin (1858), pp. 87–90. Richards (2017) is now much the most comprehensive study.

differentiated. Natural selection is confirmed in its powers as causing both adaptation and progress, and the analogy with artificial selection is confirmed in its value in evidencing those powers.[38] When the time came to write the *Origin*, and prepare to give prominence in print to these doctrines about divergence, there was no need to rethink the selection analogy, only to draw new comparisons between the consequences of the struggle and the consequences of horse breeding and other such artificial breeding practices.

Even a brief look at this new divergence theorising in the 1850s raises a question fruitfully discussed by Darwin buffs over recent decades. How much did his thinking about natural selection change between the early 1840s and late 1850s? On two clusters of topics shifts happened: variation in the wild and circumstantial influences there. Thanks especially to his barnacle studies in the 1850s Darwin became convinced that there was far more hereditary variation, arising in all species at all times, than he had previously thought. And his divergence theorising itself led him to think more ecologically than before about competition and struggle. The term itself is an allowable anachronism as labelling the theorising found in many authors writing about these two subjects under the 'economy of nature' rubric. Lyell's biogeographical theorising about species extinctions was a prime exemplar for Darwin, who in the 1850s went on to put less emphasis on soil and climate and the like as influences, and more on changes in how plants and animals interacted intra- and inter-specifically in influencing each other's lives and deaths; and in how consequently they caused both hereditary variation as material for selection, and changes in what selection was favouring at different times and places. Accordingly he thought geographical isolation less necessary for species formation and diversifying selection without isolation more effective. In these senses his theorising became even more ecological than before. The integrations between his selection theory and his breeding analogy were strengthened and amplified by these shifts, but neither theory nor analogy were altered in structure or function.[39]

Familiarly enough, Darwin's writing of the *Origin*, as an abstract of the much larger unfinished *Natural Selection*, was prompted by Wallace sending him, in 1858, from what the British then called the Malay Archipelago, a short essay which, both men soon agreed, presented a theory very like

---

[38] Ospovat (1981) and Kohn (2009). Partridge (2018) exaggerates the differences between 1844 and 1859.

[39] Kohn (2009).

Darwin's theory of natural selection. This coincidence has prompted recently some misguided claims that Darwin stole ideas, especially on divergence, from Wallace; and also some cogent but mistaken claims that the two men's theories are less alike than they themselves thought. One such claim is that Wallace opposed any comparisons of wild and domestic life and so any selection analogy such as Darwin's. This mistake can clarify the analogy itself. Yes, Wallace's essay does oppose arguments invoking comparisons between domestic and wild varieties, but, no, this opposition in 1858 did not prevent him welcoming the selection analogy, most probably on first reading Darwin's writings in 1859–1860 and for the rest of his very long life.

Wallace's essay counters an argument against the transmutation of species. When domestic varieties go feral they sooner or later revert; therefore, the argument went, any wild varieties of wild species may be presumed to do this too. Such varieties would then be unstable, thus ensuring that species are stable. Wallace counters by insisting that domestic animals are so very different from wild animals that no such conclusions about wild species can be inferred from generalisations about domestic animals, whether they are living on the farm or in the woods. The decisive difference is that wild animal life – and he is not including humans, who are, he will later make explicit, social and sympathetic – is a competitive struggle for existence by strong, energetic, self-helping individuals finding food and avoiding predators unaided by others of their own or any other species; while domestic animals are feeble, effete and entirely dependent on human help with feeding, protection from predation and harsh weather. Any wild varieties whose individuals are superior strugglers will replace their common parental species, and this adaptive divergent and progressive replacing will be reiterated without limit in the long run, while domestic varieties will only survive in the wild by reverting to the characters of their wild ancestors.[40]

Wallace's argument can easily lead to a misreading of a couple of sentences about domestic varieties early in the *Origin*'s first chapter, referring to a 'statement often made by naturalists' that domestic varieties on running wild revert to the character of their original stocks. From this, Darwin says, these naturalists argue that no inferences 'can be drawn from domestic races to species' in the wild. Is Darwin countering Wallace here? No, and for two reasons. First, Darwin knew all too well that Wallace was not a member of any plural anti-transmutationist naturalist opposition that

[40] Wallace (1858).

needed discrediting. Second, these anti-transmutationist naturalists were defending the unchanging stability of species by pointing to the reversion of domestic varieties gone feral, and by inferring that wild varieties of wild species will also revert, and so not lead to any changes of one species into another. Now, this anti-transmutationist argument is the very one countered by Wallace in 1858. For he argues that domestic varieties are so ill-fitted to the rigours of life in the wild that they can only survive out there by reverting; while wild varieties arising in wild species will include some that are even better at struggling for existence than the parent species stock, confirming therefore that effete domestic varieties and rugged wild varieties are so unalike that no inferences, about the unchanging stability of wild species, can be drawn from the fate of domestic varieties in the wild.[41]

This countering by Wallace of this anti-transmutationist argument for species stability leaves him well able to join Darwin, in insisting that conclusions from varieties made by artificial selective breeding on the farm can support conclusions about wild varieties in the wild. Wallace's annotations on his copy of Darwin's first edition indicate that he thought carefully about this passage. But there is no sign that he thought Darwin was opposing his views rather than those that he, Wallace, had been countering in his essay.[42]

Quite generally Wallace's 1858 account of wild animal life and its consequences manifestly matches Darwin's very closely. On domestic animal life there is obviously one decisive difference: artificial selective breeding has no place in Wallace's account and goes unmentioned. But he could easily add it in, especially as it was one more instance of domestic animal life being unlike wild life in its dependence on man's intervening care and control. Moreover, because Darwin's analogy is a relational proportionality, the two relata, the struggle in the wild and the stockbreeding practices on the farm, can be utterly unalike; for it is not these relata that have to resemble each other but their causal relations to what they act upon. Unsurprisingly then, Wallace's annotations in his copy of the *Origin*'s first edition, nowhere reject the selection analogy, and the analogy features often in his writings after his return to England in 1862, and is introduced and defended in his book *Darwinism* in 1889 in almost exactly the wording Darwin used in March 1839. What then has misled historians into assuming that Wallace was a reluctant and tardy convert to the analogy? Probably a confusion between variation with selection and variation without.[43]

---

[41] Wallace (1858); Darwin (1859), p. 14.   [42] Bedall (1988), pp. 270–275.   [43] Ibid.

Darwin does argue as follows: domestic animals vary hereditarily because of changes in circumstantial influences such as climate and nutrition; and wild animals, geology shows, are likewise subject to such circumstantial influences, so they are inferably varying hereditarily. This inference does not concern selective breeding, natural or artificial. Hereditary variation in the wild is being evidenced by Darwin not directly by observing it in animals living in the wild or in museum specimens, but indirectly by extending a causal generalisation from farms today to the forests and prairies in past, present and future ages. To this indirect evidencing Wallace responded by urging himself and Darwin to do more direct observational confirming of variation in the wild, so reducing indirect evidential dependence on inferences from domestic animal and plant variation. But this response to this evidential issue assumes and implies no wariness or misgivings about the selection analogy and its evidential duties and virtues.[44]

Wallace made, in 1866 correspondence, another recommendation that Darwin substitute Herbert Spencer's 'survival of the fittest' for 'natural selection' to keep unsophisticated readers from misconstruing its metaphorical import. Both men did some substituting, but both continued writing of 'natural selection'.[45]

Even this brief glance at Wallace and Darwin, before and after they agreed about their remarkable agreement, can confirm that neither of them first arrived at the theory of natural selection by arriving at an analogy which was somehow the theory itself. For both of them, the selection analogy was always supporting without ever being or becoming the theory. The *Origin* was, Darwin said, one long argument. This was true of a book distilled from two decades of consistent and persistent argumentation from 1839 on. No one has ever read it, the selection analogy included, with more attention, admiration, comprehension and agreement than Wallace.

[44] See the discussion in Chapter 8.   [45] Bedall (1988), pp. 275–277.

CHAPTER 4

The 'One Long Argument' of the Origin

Our previous chapter's account of the first two decades of Darwin's theorising about natural selection shows how the *Origin* conforms to the vera causa compositional and evidential ideals. It shows too how the selection analogy conforms to the ancient causal proportionality ideal. And so it shows how these three ideals all conjoin to instruct the arguments for those two big ideas: the tree of life and natural selection.

The Elements and Their Integrations

The book's first edition begins by explaining how it abstracts the much bigger and incomplete *Natural Selection* treatise, and how the new book's fourteen chapters (see Table 4.1) proceed. In opening the last chapter, Darwin calls the *Origin* 'one long argument', before reflecting on the argument's main steps and wider implications across and beyond the sciences of nature. In the thirteen chapters before this, the relationship between the first four (I–IV) and final five (IX–XIII) is most decisive for the structure and strategy of the long argument. For, in accord with the vera causa ideals, the argument presents, as the *Sketch* and *Essay* did, three evidential cases on behalf of natural selection: cases for its existence and adequacy (I–IV) and for its responsibility (IX–XIII). Of the intervening four chapters (V–VIII), one (V) supplements the opening two (I–II) on variation on the farm and in the wild; three (VI–VIII) then counter objections to the adequacy case made in chapter IV. The book's punchline is what those later five chapters (IX–XIII) argue for: natural selection (mostly) did it; that is Darwin's theory. By evidencing the existence and adequacy of natural selection, the first four chapters enable those later five to argue for this responsibility conclusion, and so for the greater probability of this theory over others. The first four assemble the explanatory resources, the laws and causes, deployed in the explanatory successes

Table 4.1 *List of chapters in Darwin's* Origin of Species *(1859)*

Introduction	
I	Variation Under Domestication
II	Variation Under Nature
III	Struggle for Existence
IV	Natural Selection
V	Laws of Variation
VI	Difficulties on Theory
VII	Instinct
VIII	Hybridism
IX	On the Imperfection of the Geological Record
X	On the Geological Succession of Organic Beings
XI	Geographical Distribution
XII	Geographical Distribution – *continued*
XIII	Mutual Affinities of Organic Beings: Morphology: Embryology: Rudimentary Organs
XIV	Recapitulation and Conclusion

articulated in the final five, the successes indirectly evidencing responsibility.

This rationale for the twofold division is not as manifest as in the *Sketch* and *Essay*, mainly because Darwin's four intervening chapters (V–VIII) relate in several different ways to the earlier four and later five. But the selection analogy is functioning in the *Origin* as it does in those ancestor texts. Writing in May 1863 to the botanist George Bentham (nephew of the more famous Jeremy, and author when young of an unusually innovative book on logic including criticism of Whately's *Elements*), Darwin insisted that

> the belief in natural selection must at present be grounded entirely on general considerations. (1) on its being a vera causa, from the struggle for existence; & the certain geological fact that species do somehow change (2) from the analogy of change under domestication by man's selection. (3) & chiefly from this view connecting under an intelligible point of view a host of facts.[1]

By contrast with these general considerations, Darwin says that when descending to details no one species can be shown to have changed, nor

---

[1] Burkhardt et al. (1985–), *11*, p. 433. For comprehensive and detailed guides to the *Origin*, see Costa (2009); Reznick (2010); Hodge (2013a); Ruse and Richards (2009). For further elucidation of the vera causa ideal and the structure and strategy of the *Origin*, see Hodge (1977). The present analysis corrects some mistakes made there.

that the changes are beneficial; nor why some species have changed while others have not. This contrast may seem to concede more than does the *Origin*, which is famously full of details about particular species in the wild and varieties in gardens, on farms and in pigeon lofts. However, working within this concession, the book's argument has all these details supporting general grounds for accepting the theoretical theses, rather than providing observed instances of branching descent or of natural selection in action.

This invaluable short guide given to Bentham confirms that relating the selection analogy to the vera causa ideals must be an exegetical priority. Darwin's claim that natural selection is a vera causa invokes the Reidian sense of this term, whereby a true cause is a real, known, existing cause whose existence is evidenced, is known, independently of the facts it is to explain. The appeal to geology and the struggle for life refers to the *Origin*'s chapters II–IV, which argue that geology reveals wild species living in continually changing conditions that, observations of domestic animals show, must be causing inherited variation. Crucially, then, the observed existence of abundant hereditary variation on the farm evidences the existence of the less abundant hereditary variation in the wild, on the assumption that the same causes are at work in the wild albeit less powerfully. But the known existence of selection on the farm does not evidence the real existence of selection in the wild; so it does not establish that natural selection is a vera causa. It is the existence of the struggle for life, together with wild hereditary variation, which evidences, because it causally entails, the real existence of natural selection there.

The second general consideration, the analogical comparing and contrasting of nature's selection with man's, establishes what nature's can do in its much longer run. So much then for the book's first four chapters. The third general consideration refers to the final five (IX–XIII) where natural selection is shown to explain many factual generalisations from palaeontology, biogeography, taxonomy and morphology.

Darwin's short guide alerts us to a tempting but misleading exegesis that would have Darwin using the selection analogy to evidence natural selection as a real, existing, true cause, and so following Herschel's teaching that a vera causa can acquire this status from analogical justification. It is time to reverse this exegetical argument: since Darwin did not argue that natural selection is a true cause because it is analogous to artificial selection, he was not following Herschel in conforming his analogical reasonings to the vera causa ideals.[2] To make the lessons from the short guide even more explicit,

---

[2] These issues are most recently treated in Pence (2018).

as the *Origin* conforms to them, those ideals required an explanatory cause to be evidenced in three independent ways: as to its existence, its adequacy and its responsibility. Darwin's selection analogy contributes directly to his long argument's meeting the second of these desiderata, and indirectly to meeting the third, but not the first, nor then to the evidencing of natural selection as a true, an existent cause, a vera causa: hence, the distinguishing and ordering of the three general considerations in the short guide for Bentham, a guide Darwin could have composed two decades earlier, had he not been withholding the ancestor texts from almost any circulation, even among fellow naturalists who were friends.

Darwin's short guide lets us see that, although there are three clusters of chapters (I–IV, V–VIII and IX–XIII), this tripartition does not coincide with his threefold distinction of general evidential considerations. To begin to see why, consider next the first four chapters.

The First of Three Chapter Clusters (I–IV)

To recap and preview, the opening chapter, on variation in domesticated animals and plants, discusses variation itself and then hereditary variation as accumulated by the art of selective breeding; the second chapter concerns variation in the wild but not selection out there; the third introduces the struggle for existence in nature, and indicates briefly how it causes in wild species a natural selective breeding, a process comparable in kind to man's selection; then the fourth argues that due to its powerful cause, the struggle, natural selection is able to produce unlimited adaptive diversifications.

So far so familiar; and now for some further recapping and previewing. There are not merely appeals to factual generalisations about hereditary variation on the farm or in the wild but arguments invoking the causes of those tendencies and processes. All four chapters emphasise that explicit contrast: the tendency to hereditary variation in domesticated species is greater than in the wild; but with selection it is the other way round: nature's selection so vastly exceeds in power man's as to more than compensate for this lesser variation in the wild. This variation on the farm and in the wild has the same causes: changes in conditions, of soil, nutrition, weather and so on that disrupt sexual and asexual reproductions which, in entirely unchanging conditions, would yield offspring exactly like the parents. However, these causes of variation are effective to a higher degree under domestication than in nature. By contrast the causes of selection under domestication and in nature are entirely unalike; although the selective consequences of these unlike causes for survival and leaving offspring are the same in kind though not in degree.

So, the first causal theme in the opening chapter (I) is that, in domestic species, the abundant inherited variation is caused especially by influences on the parents affecting their reproductive elements prior to the conceiving of their offspring, and also by the effects of habits, or indeed by the direct action of diet changes and the like; while the second causal theme is that selection by man, rather than crossing or inbreeding, has been the main means whereby this inherited variation has been accumulated over successive generations so as to make varieties or breeds serving man's uses or fancies. Conspicuously there is no sustained emphasis here on hereditary variation due to chance, to the action, that is, of small, hidden, unknown prenatal causes that produce useless and unwanted as well as useful and wanted variations; but the implication is that selection is efficacious in working with any hereditary variations whether due to chance or not. This efficacy is evident in 'methodical selection' when a breeder works deliberately to make a variety to fit particular needs or wishes; and even more in the 'unconscious selection' resulting from the practice of breeding from the best individuals over many generations with no conscious intention of changing the whole breed.[3]

There is one causal theme belonging in chapter II but explicit only in the recapitulation at the book's end: geology shows that every region has been and still is continually undergoing physical and faunal and floral changes, so animals and plants are at all times caused to vary under nature just as they are in changing conditions under domestication. This second chapter does explicitly argue that species in larger genera usually have more varieties than those in smaller genera, because of the greater variability in wider ranging groups exposed to more varied conditions. Here, Darwin interprets varieties as incipient species and species as well-marked varieties differing in degree but not in kind from varieties.

Chapter III emphasises that there is always in the wild a competitive struggle to survive and reproduce owing to the tendency of all species to increase their numbers, and to the checking of those potential increases by such checks as predation and food limitations. The principal causal theme of the fourth chapter is anticipated here, when Darwin identifies this struggle as the cause of persistent and consistent, and so selective and adaptive, accumulation of hereditary variation.

---

[3] Darwin (1859), pp. 33–35.

The fourth chapter itself starts with the greater power of nature's over man's selection due its being more prolonged, precise and comprehensive in selecting over eons among all those very slight variations making for small, but in the long run decisive, differences in chances of success or failure in the competitive struggle to survive and reproduce. Complementing this ordinary natural selection is sexual selection in competition to win mates through male combat or female choice with arms (stags' antlers) or charms (peacocks' tails). In the middle four chapters, sexual selection will be integrated with generalisations about variation in secondary sexual characters and will be seen to enhance the causal adequacy of selection generally, especially in causing features too disadvantageous in the struggle for life to be due to natural selection. However, in the later cluster of five chapters (IX–XIII) Darwin will find no explanatory work for sexual selection. This special complement to ordinary natural selection, as it is in the *Origin*, has to wait until 1871 and the *Descent of Man*, to come fully and publicly into its own as an explanatory resource not least in accounting for human racial diversity.

The end of chapter IV is dominated by the principle of divergence: structural and functional specialisation is usually advantageous in life's struggle; so over eons natural selection causes, reliably if not invariably, structural and functional divergences in favouring diverse adaptive specialisations, so causing branching descents among the more specialised winning species and terminal extinctions among less specialised losers. Such increased structural and functional differentiation in animal and plant organisation constitutes progress. As a reliable cause of adaptation natural selection is no less a reliable cause of progress.

This fourth chapter of Darwin's *Origin* is therefore decisive, as the site where the marshalling of vera causa kosher causal–explanatory resources is completed with the selection analogy's support for the causal adequacy of natural selection. We will revisit this site after finishing our tour of the whole book.

There are places where Darwin calls a cause other than natural selection a vera causa. Common descent is so identified more than once, most instructively in contrasting the theory of common descent with the 'the ordinary view' of species as each independently created. On this view, Darwin says, the similarity in the enlarged stems of three turnip species would have to be attributed 'not to the *vera causa* of community of descent, and a consequent tendency to vary in a like manner, but to three separate yet closely related acts of creation.' It is, however, only his

conceiving of natural selection as a real, true, known, existing cause that
informs his structuring of the long argument.[4]

Community of descent as a vera causa. has its place in the causal–
explanatory complex marshalled in the course of chapter IV, the complex
defended and deployed throughout the rest of the long argument. In IV
and thereafter, natural selection is, like domestic selection, branching and
divergent, and can be because descents, natural and domestic, are common
and so branching and divergent.

The Second Cluster (V–VIII) and the Third (IX–XIII)

What follows IV is that middle miscellany of four chapters, opening with
one (V) on the laws of variation, and with Darwin saying that he has
spoken of some variations being due to chance, so improperly implying
that not all variations are due to lawful causes. In fact he has talked only
infrequently about chance variations, and the larger aim of this chapter is
to secure a unification thesis. Variations in domestic and wild plants and
animals all conform to the same laws; organs developed to an extreme
degree in some organisms being, for example, very variable also in their
more normal close relations. Again, the laws of variation are the same for
species as for varieties: structures varying between species vary similarly
within species. So this chapter contributes to the second chapter's thesis on
variation in nature (II): that species and varieties differ in degree but not in
kind; species being well-marked varieties and varieties incipient species.
And this thesis supports the argument running throughout the book: that
the causes of species, as well-marked varieties, differ in degree but not in
kind from the causes of both wild and domestic varieties.

Next, chapter VI counters reasons for thinking natural selection incapable of forming new species from old, because some have features that
selection cannot produce, especially organs of extreme complexity and
perfection such as the eye. Darwin's countering appeals to the existence
today of a graduated array of useful organs, from the eye on down to
simple structures conferring mere sensitivity to light, an array making it
conceivable that eyes could have been produced gradually over eons by
natural selection. Chapter VII deploys the same strategy in discussing the
complexity and perfection in instincts such as bees exhibit in building their
combs. In explaining how sterile neuter insects could owe their instincts to
natural selection when they do not breed over successive generations, he

[4] Ibid., p. 159.

argues that if selection is admitted to take place among families as well as individuals the difficulty can be overcome. He implies that this admission entails no significant amending of his theory and his selection analogy, as farmers have improved the quality of castrated steers by consistently breeding from the parents of the best.

Chapter VIII deals with the objection that species are unlike varieties in their inability when crossed to produce fertile hybrid offspring; and that this inter-sterility permanently ensures the distinctness of species and is not a property of species that natural selection could produce. Darwin here disputes the view that all and only species and no varieties are inter-sterile, for inter-sterility is not always either completely present or totally absent in species or varieties but comes in degrees. And he argues that, while not directly due to natural selection, inter-sterility can be indirectly so, because it is a gradual, incidental consequence of those adaptive divergences in hereditary constitutions produced by natural selection over many generations. Here, too, he sees no need to amend the theory or to question the support for the adequacy case provided by the selection analogy.

The following chapter (IX), although often associated with the previous four, really belongs with the later four (X–XIII), those displaying the theory's explanatory virtue, and so providing evidence for natural selection having been responsible for producing the extant species living today and the extinct species commemorated as fossils. This chapter argues that the fossil record is not a complete and reliable record of sporadic, sudden and jumpy exchanges of new species for old; rather it is a patchy, gappy, intermittent, damaged, fragmentary little-studied record of what were gradual transitional changes in species and their conditions of life. Properly interpreted, this rocky record presents no insuperable difficulty for the view that those changes were slowly and smoothly wrought in gradual branching descents by means of natural selection, descents analogous to those lesser branching adaptive diversifications wrought by man's selection in much briefer recent times.

So, as dealing with a difficulty, chapter IX may seem to belong in the middle cluster, but as dealing not with adequacy, but with responsibility case difficulties, it really belongs with the four chapters (X–XIII) presenting positive responsibility arguments from explanatory successes. Not only do those four begin with the second of two chapters devoted to geological issues, that second geological chapter (X) ends with a single summary comprehending them both (IX–X). Moreover, it is followed by two chapters (XI–XII) on biogeography, again with the first taking care of difficulties, and the second mostly presenting explanatory successes and ending by summarising both. But, just when the reader expects this pairing pattern to reappear, Darwin completes the book's whole argument with a single chapter (XIII), albeit almost the longest, on classification, morphology and embryology. Not that the pairing pattern is kept to tidily. If defensive counterings of potential defeats are distinguished from winning constructions of actual successes, the second geological (X) and the second geographical chapter (XII) are both seen to have some of each; while XIII has no defendings, only triumphs. But despite this expositional untidiness, all five chapters (IX–XIII) taken together do make the responsibility case for branching natural selection, by arguing that there are no insuperable difficulties for this case, and plenty of explanatory successes favouring it.

With these nerdy morsels ingested, let exegesis recommence; and first by previewing complications in how Darwin conducts his positive responsibility case. Yes, there is branching descent by means of natural selection; often branching descent, and so common ancestry, is in play, but natural selection is not. When natural selection is in play, common ancestry is too, because, although natural selection can cause unbranching descent, Darwin thinks it does not commonly do so. Descent is usually branching and also gradual and irregular: gradual in not being jumpy, and irregular in not always branching into three or five or any other privileged number of sub-branches.

Branching change caused by natural selection is directed by circumstantial contingencies, and results either in adaptation to these contingencies or in extinctions. But could any other processes than natural selection cause gradual irregular, circumstantially contingent adaptive change? Darwin seems to assume not; but one obvious candidate is never explicitly ruled out: the causation later dubbed Lamarckian, which is driven by the inheritance of non-chance variations arising in adaptive responses to circumstantial contingencies, causation which Darwin explicitly accepts is sometimes at work in wild as in domestic animals and plants. More than anything, arriving at his selection analogy late in 1838 had soon led Darwin to downplay, even to marginalise, Lamarckian causation. The issue of causal–explanatory alternatives also concerns branching descent. Can the factual generalisations Darwin explains as due to branching descent have no other cause and be evidence for no other theory? Darwin is often saying they could not, so that these explanatory successes are achievable if and only if branching descent is accepted.

On the representation of branching descent itself, the tree of life metaphor appears to fit such descent exactly; and yet a question arises as to how that fitting works. The branching diagram in IV seems to be clearly a species propagation diagram like one early in *Notebook B*.[5] In representations of arboriform species propagations, the existence, the duration, of a species is represented by a line. And the increasing character gap between co-descended diverging species is represented by their life lines diverging. A line can split into two or more lines representing the lives of so many descendent species, each line being one species wide. A genus of closely similar, very recently co-descended species is represented by two or more of these species lines, and a family by equally thin lines representing the many species of several genera descending from a more remote single ancestral species. And so on with wider and wider classificatory groupings of species descending from more and more remote singular common ancestral species; all of these ancestral and descendent species durations and splittings and divergings are represented by extensions and branchings of equally thin lines.

We will have to ask also, then, why branching species propagation delineations have causal–explanatory priority for Darwin. These delineations do because they reconstruct causal processes while a taxonomy presents the groups-within-groups classificatory relations among the products of these processes. The nested groups' production was not proceeding as it did because they would become classifiable as they are; but, rather, they are so classifiable because so produced. Crucially, a natural classification is, for Darwin, instructed by a genealogy that is constructed from species propagation tree evidence for branching natural selection as modelled by artificial selection. And this instruction accords with those causal generalisations that are common to the artificial model and its natural target alike. The principal generalisation is about non-adaptive resemblances. The resemblances between the pentadactyl structure of a man's grasping hand and of a whale's swimming flipper are non-adaptive, because they are not explicable as common adaptations to common circumstances and ways of life. Such resemblances are, rather, produced by and evidence for common ancestry, whereas the adaptive differences between the hand and the flipper are due to and so evidence for divergent natural selection over the eons after the ancestral species had this common structure. As for men and whales among mammal species, so for racehorses and drafthorses among equine breeds: common ancestry explains the resemblances not ascribable to common adaptations; and divergent selections explain the adaptive differences.[6]

[5] Ibid., pp. 116–118.   [6] Ibid., p. 129 and pp. 434–438.

These species propagation tree themes are introduced in ending IV, the natural selection chapter, and are integrated there with classificatory, taxonomic themes. The integrating of these themes is founded in an integrating of those two big ideas: common ancestry with branching descent and natural selection. All these integrations are completed in XIII. Only by looking at their initiation in IV can we understand their completion late in the book. All the chapters after IV are sequels, but none are more sequels than others. They do serve as sequels in various ways, although one element is constant: the structure and function of the selection analogy, as introduced in IV.

Selection: From Chapters I to IV

Revisiting the natural selection chapter (IV), we look again at what comes before and after. From chapter I on there are two simplifications. Not only are the laws and causes of hereditary variation the same in the woods as on the farm; on artificial selection Darwin forewent full coverage of breeders' theories and practices, by largely ignoring crossing and inbreeding. His giving priority to selection began with the Malthusian moments in late 1838. The selection analogy was fitted to Malthusian wedging and sorting. Absent any Malthusian crossing or inbreeding, this rationale for this narrowing endured in all Darwin wrote thereafter.

The integrating of propagational and taxonomic relations begins in the closing summary of chapter IV, but, long before this, another integration is preparing the ground. For, from chapter I on, Darwin is integrating ancestor–descendant causal relations, genealogical relations in his terms, and selectional causal relations. Even before discussing artificial selection in chapter I, he is bringing his detailed account of domestic pigeons to bear on this integration in arguing, against many pigeon breeders, that all the diverse domestic pigeon varieties have descended over centuries from a single wild ancestral species. Although a polygenist about dog varieties, he is a monogenist about pigeons. Given that selection has been the main cause of divergent descent from the single common ancestral pigeon species, the diversity among pigeon varieties is evidence for the power of man's selective breeding. Conversely the power of selective breeding in diversifying other domestic species is evidence for the possibility of a single pigeon ancestry.

Of the two kinds of artificial selection, methodical selection comes first and unconscious selection is characterised by comparison and contrast. Methodical selection might have been dubbed deliberate or intentional,

and unconscious called undeliberate or unintentional. Two distinctions
help: mental versus physical actions, and necessary versus sufficient causal
conditions. A decision to castrate some inferior young bulls is a mental
action, but may be a necessary condition for the physical action, the
castrating. With methodical selection the mental actions may include
deciding to improve a breed of sheep, say, or even to make a new and
superior breed. And the physical actions necessary and sufficient for this
outcome are not just deliberate, but undertaken with that remote future
outcome consciously in mind. Not so with unconscious selection. Here
the best dogs, say, are deliberately kept and the rest culled; but this is done
with no conscious intention of improving the breed over future generations. The limited intention to keep the best is necessary for this future
outcome; but the outcome is an unintended consequence of acting physically on that limited intention.

What makes any artificial selection not natural is that the causation is
both intentional and interventional in ways ensuring outcomes fitted to
men's uses and fancies. It is interventional because what the intentional
human actions are making happen physically would not happen were the
natural processes left to themselves. For Darwin, the making of domestic
varieties fitting men's uses and fancies is no lucky accident, but an obvious
consequence of the sustained succession of mental-intentional and
physical-interventional human actions producing them. Artificial selection
is what a human stockbreeder does to and with hereditary variation in
nonhuman domestic species. Nature's selecting is natural because it is what
the struggle for existence does to and with hereditary variation in wild
species, in ensuring that the resulting varieties are eventually fitted to the
animals' and plants' own ends, of survival and reproduction – which is
again no lucky accident, but manifestly a consequence of the physical
actions producing them. Some struggling in the wild may or may not be
effective because of intentional inter-specific interventions; lions may or
may not intervene intentionally in the lives of deer. All predation and
grazing entails such interventions; but although inter-specific, these interventions are independent of human agency and serve no human end, and
the interventions of floods and blizzards are no less effective in being
obviously unintentional. Likewise too with sexual selection, as modelled
by the artificial selection practised by a human breeder who desires and so
selects more and more handsome bantam cocks; he acts intentionally and
interventionally, so ensuring an outcome fitting his fancies. The struggle
for mates in the wild may or may not be struggles of consciously intending
animals; but, whether through competitive armed combat or competitive

charming display, the fitting of their descendants for future reproductive success depends on the physical consequences physical differences have for competitive intra-specific, intra-sexual winning and losing, and not on intentional interventions by any humans. As for sexual selection among humans, this is natural in not being an intentional intervention by humans in the reproductive life of another species, and in being, like sexual selection among peacocks, analogically comparable to the artificial sexual selection of the most handsome bantam cocks. Some imaginary human eugenic breeding policies and practices may not be easily categorised as natural or artificial, but none such feature as models or as targets in Darwin's analogical articulation of his theory.[7]

Darwin is usually clear enough as to what is model and what target, and what is artificial and what natural. Any selective breeding situation, natural or artificial, in the wild or on the farm, could be a model for any other selective breeding target situation, providing Darwin can learn from the better known model about the lesser known target; but, his model situations are almost invariably artificial because they are better known than his natural target situations; so he is usually inferring and learning more about nature from art than about art from nature.

On artificial selection generally, Darwin most decisively insists that if selection were only a separating and breeding of some° distinct variety, it would be too obvious to dwell on; rather what is important is the great effect produced by the accumulation in one direction, over successive generations, of differences inappreciable to an inexpert eye, even indeed to his own. Consistent, persistent, minutely discriminating breeding producing gradual accumulative change is the essence of this 'principle' of selection.

This view of this art grounds Darwin's analysis of the circumstances favourable and unfavourable to its efficacy. Favourable are plentiful hereditary variability, and large flocks or herds as offering more variation; most obviously favourable too is expert close attention to tiny differences. Unfavourable are any circumstances conducive to the crossing of favoured with unfavoured individuals and varieties. Social and economic preferences are revealed here. Enclosure, private ownership by the wealthy of fenced land formerly worked as unfenced commons by poorer folk, entails less crossing than before; while savage and nomadic peoples' life-styles facilitate crossing among their domesticated animals. Pigeons who pair for life are more easily fitted by selective breeding to men's uses and fancies than are

---

[7] Ibid., pp. 87–90.

promiscuous cats. As familiar human causation, selection is easily understood and assessed as indisputably efficacious; and, although acknowledged
as a pre-eminently British triumph of recent decades, has long been
recognised in theory and practice across the civilised world. The origins
and subsequent histories of the breeds it has produced are rarely recorded
and so little known; but Darwin is not troubled by this ignorance,
for we would expect it given all we know, and especially what we need
to know about the power of artificial selection as evidence for the power of
natural selection.[8]

The table of contents at the opening of chapter IV begins with three
invocations of power: the power of natural selection compared with man's;
its power in affecting unimportant characters, and its power at all ages and
for both sexes. The relation of chapters I–IV derives from this power-
driven analogical reasoning from artificial to natural selection, reasoning
that may look inconsistent but is not. In chapter I weight is put on the
great power of variation with artificial selection – as contrasted with
variation on its own – to improve and make adaptive breeds; whereas in
chapter IV what is most emphasised is the manifestly lesser power of this
artificial selection as contrasted with nature's selection. There is no inconsistency. The entire argument is establishing the limitless power of natural
selection, by first talking up the power of artificial selection, and then
talking up the power of natural selection as superior not to a feeble artificial
power but to a very effective one. For Darwin is preparing to argue a
fortiori, from the stronger. His a fortiori arguments are not arguments
making stronger the reason for accepting some conclusion; they are
arguments from the greater strength or power of one cause compared with
another, and so to the greater extent of one lot of effects over another. His
argument a fortiori is an argument not to but from the analogical four-
term causal–relational proportion wherein the struggle for existence is to
wild animal variation as the stockbreeder is to domestic animal variation.
In establishing the adequacy of natural selection to cause the adaptive
diversification of species, the a fortiori reasoning allows Darwin's long
argument to conform to the vera causa evidential ideal.[9]

A Comprehensive Causal–Explanatory Complex

The main business of chapter IV is to marshal the comprehensive causal–
explanatory complex defended and deployed throughout the rest of the

[8] Ibid., pp. 29–35 and 41.   [9] See the analysis of the a fortiori arguments in our Chapter 5.

book. This marshalling integrates natural selection with branching and divergent descent and with species extinctions. The selection analogy contributes to this integrating, and is invoked throughout the rest of the book. The subsequent argumentation begins defensively with chapters VI through VIII countering objections, as do chapters IX and XI, while chapters X, XII and XIII are no less dominantly on the offensive. The selection analogy serves defensive and offensive argumentation alike.

At the core of the comprehensive complex are the principle of selection, as defined by stockbreeders and conformed to in their practices, and the principle of natural selection, so named by Darwin to mark its relation to man's power of selection. Both principles are principles effective causally because of their own respective causes: the stockbreeders' judgements and interventions and the struggle for existence. Back in chapter I Darwin has emphasised that it is evident to the stockbreeders themselves why their expert decisions and actions have the powers and so the effects they do. In chapter III, he emphasises that it can be no less evident why the struggle for existence should have effects similar in kind but much greater in degree, once one understands the causing of that struggle. The struggle for existence follows inevitably, he says, from the high rates of potential population increase common to all plants and animals, and more precisely from the difference between these potential increases and the much lower rates of increase, often zero or negative, actually present in the wild. This difference is due to the checks on those potential increases, from predation or from food shortages and the like. With this excess fertility, and so with more individuals being produced than can survive, there are struggles for existence, struggles between one individual and another or with the physical conditions of life. Malthusian excess fertility causes the struggles, and because so caused the struggles can cause selection. For, thanks to these struggles, any hereditary variation, however slight and however caused, if at all advantageous to an individual in its relations with others or to physical nature, will enhance that individual's chances of survival and will be inherited by its offspring who will have their chances enhanced too. The struggles due to excess fertility causally entail a selective breeding in nature like that on the farm.

This core causal reasoning brings out once more how Darwin is working in his book, as in his earliest notebook selection theorising, with two quite distinct causal concepts of chance: chance as accident, an ancient concept, and chance as a probability, with a value between zero and one, an equally commonplace but more modern concept. Some hereditary variation is chancy; it is what ancient writers called accidental or fortuitous in its causation and was still called that in Darwin's day and in his notebook theorising. Now consider not the cause of this variation but its effects, its consequences. Some of this chance variation may be advantageous in the struggle for existence. If it is, this consequence, the advantage, is not accidental at all, but quite the opposite. It is no accident, no matter of mere luck, that in a species of bugs green individuals will be surviving more often than red ones when all are living on green vegetation and mostly dying from predation by colour-sighted birds. Being green rather than red, this advantageous colour difference, is positively probabilistically causally relevant to survival and reproduction in these circumstances. The green bugs' better chance, their higher probability, of survival and reproduction is non-fortuitous. And likewise for the lower probability of the red ones surviving. Natural selection is what happens in the wild when there is causally non-fortuitous differential survival and reproduction of hereditary variants which may have been generated, caused, accidentally, fortuitously. The aptness of this explication of Darwin's causal reasoning is evident because if the birds are all colour-blind then any greater surviving of either red or green bugs would be causally fortuitous: survival of the luckier not of the fitter. The explication applies equally to artificial selection. Due to accidental, chance, fortuitous hereditary variation, some foals may be swifter than others and, if living in the causal circumstances deliberately arranged by a stockbreeder on a racehorse stud farm, these foals will be caused to have enhanced chances, higher probabilities, of survival and reproduction, whereas in other stockbreeding circumstances, where only superior strength is favoured, it may be a matter of luck whether the swifter are surviving more or less often than the slower.

These Malthusian themes and their selection analogy articulations are obviously invoked throughout the chapters after IV. But that chapter is especially decisive because Darwin elaborates there his comprehensive causal–explanatory complex concerning natural selection's causing of inter-specific branching and divergent descent and extinctions, just as artificial selection does intra-specifically. This elaboration confirms that, while the book may be all about the origin of species by means of natural selection, it is no less about the species-propagational tree of life, which is also a tree of deaths, extinctions, as produced by natural selection; and that it is not the origin of species but this tree's production that is the consummating topic at the closing of chapter IV, the chapter devoted to natural as analogous to artificial selection; and again in the penultimate chapter XIII at the closing of the long argument. Commentators often emphasise, correctly enough, that the tree of life and natural selection are

two distinct theses, and that many of Darwin's readers have been per-
suaded by Darwin's case for the first while not going on to embrace the
second. But the book's argument is designed not to segregate but to
integrate the two, and to do so often through their common invocations
of the selection analogy.

The integration is started and largely completed in the last two dozen
pages of chapter IV; but these pages naturally draw on the earlier portions
of this chapter where natural and sexual selection are compared and
contrasted with each other and with artificial selection; and the causes
and effects of nature's selection are extensively illustrated by means of
examples many of them explicitly imaginary. The emphasis throughout
the illustrations is on the consequences of the intricate and sensitive
interactions among species. As a final example, Darwin explains how a
flower species and a bee species could become adapted to each other by the
sustained selection in both species of individuals possessing favourable
variations of structure.[10]

With the action of natural selection so clarified, Darwin analyses the
circumstances favourable to its efficacy, in pages obviously cohering with
those in chapter I on the circumstances favourable to the action of artificial
selection. As preparation he has a long digression on the inter-crossing of
individuals, establishing that no animal or plant species reproduces end-
lessly by self-fertilisation, for cross-fertilisation is in the long run indispensable. The connection of this digression with the analysis of the
circumstances favourable to artificial and to natural selection is direct. In
man's conscious methodical selection a breeder selects with some definite
object in mind, and free inter-crossing will frustrate this aim entirely.
Conversely, when Darwin sums up his conclusions as to the circumstances
favourable to natural selection, he emphasises large population numbers,
wide-ranging species, severe competition, conditions varying over periods
and areas, and checks to the negative effects of inter-crossing. These
circumstances are ensured by the causes geologists, Lyellian geologists
especially, invoke in explaining how continents are transformed locally
into islands and these into continents. A large continental area of land
subject to oscillations of level will be for long periods broken into large
islands. Before this fragmentation there will be severe competition; with
fragmentation, inter-crossing will be prevented; and later, with the rising
and reforming of continuous continental land, the varieties that have now
become species will be unlikely to interbreed.[11]

[10] Darwin (1859), p. 95.   [11] Ibid., pp. 102–108.

With the securing of these conclusions comparing the circumstances favourable to natural and artificial selection, Darwin can move to the final stage in his marshalling of the causal–explanatory complex.

Extinction, Branching and Divergence

Darwin begins this final stage with two pages on species extinctions, before closing with over a dozen on branching and divergence. Natural selection causes extinctions; but it does not do so directly through its effects on the species becoming extinct, but indirectly by producing improved adaptations in other species. Any area is fully stocked with animals and plants at any time, so any increases in population in these adaptively improved species will be at the expense of decreases and hence rarity in unimproved ones. Rarity is disadvantageous, because the action of natural selection is enhanced by wide ranges and larger population numbers. Becoming rarer, some losing species eventually become terminally rare. As in nature so on the farm: extinctions of older less-favoured breeds result from the improvement by selection of new forms; very quickly sometimes, as with the longhorn cattle in Yorkshire, swept away by the shorthorns as if, in the words of 'an agricultural writer', by a murderous pestilence.[12]

Moving to divergence, Darwin again calls varieties incipient species, species in the process of formation. And he now asks how the smaller differences between varieties become the larger differences between species. As always, he reflects, he looks for light from domestic productions, finding there 'something analogous'.[13]

One man fancies pigeons with longer beaks, another prefers shorter ones; both are favouring extremes. Likewise, the breeding of light, swift horses and heavy, strong ones has established two distinct breeds, the inferior animals with intermediate characters being often neglected and largely disappearing. With man's breeding, this principle of divergence has caused steadily increasing differences as breeds have diverged from each other and from their common parent stock.

Can any 'analogous principle' apply in the wild? It can and does very effectively, because the more structurally and functionally diversified the descendants of any one species become, the more they can 'seize on many different places in the polity of nature' and so increase their numbers, which is why a plot of ground yields more herbage if sown with many species of grass, not just a few. The 'advantage of diversification' among

[12] Ibid., pp. 110–111.   [13] Ibid., p. 112.

the animals and plants in any region is like the advantage of the 'physio-logical division of labour' in the different organs of an individual body.[14]

Darwin now asks how this principle, of the benefit from divergence of character, combined with the principles of natural selection and extinction, will tend to act; and he answers in some half-dozen notoriously dense pages largely devoted to a diagram illustrating how this causal combination acts in the short and long run. Fortunately, our exegesis of his long argument does best to follow his prose, for his conclusions are illustrated but not evidentially enhanced by his diagrammatic representations, while some explicit premises are not represented at all.

Darwin's diagram is limited by what lines on paper can represent. The parallel horizontal lines mark successive periods of time from the older ages lower on the page to the younger higher up, in explicit conformity with geologists' diagrammatic conventions. A single vertical line depicts one species persisting through these ages, with no change in character that would be marked by a veering to left or right on the page. A less boring line may end without splitting and branching, going extinct without descen-dants. More consequentially a species line may split and branch without ending, with its descendent lines diverging from one another more and more in character and so eventually becoming new species lines.

Here Darwin's commentary gets more overtly causal and explanatory. Competition is most intense between structurally and functionally very similar varieties and species. Conversely, divergence is due to selection within any species favouring those varieties least like each other, such varieties being pictured at the extreme edges of small fanlike arrays of thinner intra-specific varietal lines. To get more causal and explanatory still he has to be reasoning in prose rather than pointing to his picture. Especially is this so when he addresses his leading causal–explanatory themes about winning and losing in the lives of individuals, varieties, species, genera and so on. For he invokes throughout his commentary on his diagram distinctions it does not represent: most significantly the distinction between species and supra-specific groups of wide range and large numbers and those with neither. This distinction is missing pictori-ally because in depicting the lives and so life-lines of winner species and losers, he uses indistinguishable lines. But the distinction is decisive for the thesis that the whole diagram is explicitly designed to evidence: namely, that the causes of later winning are the effects of earlier winning. Wide range and large numbers result from competitive territorial and

[14] Ibid., p. 115.

populational success, and are in turn favourable to future adaptive improvement by natural selection, all for reasons established by comparing the conditions favourable for artificial and for natural selection. Not that the diagram is inconsistent with these assertions. In showing how intra-specific variation leads to inter-specific differences, it agrees especially with Darwin's reasoning from man's artificial breeding achievements to nature's powers of adaptive diversification.

The diagram and the accompanying commentary are revealing of Darwin's views on how natural selection and the species-propagational tree of life are related, not because the diagram and commentary break new ground, but because they integrate so much that is argued for in his first four chapters, and because they dwell so minutely on the implications of his casual-explanatory complex, and because they confirm how much that complex owes to his selection analogy as it has been elaborated in those chapters.

Darwin's chapter summaries vary markedly in how they relate to the structure and content of the book as a whole. The summary for chapter IV is unique in taking up new issues treated fully in much later chapters where, as Darwin acknowledges, they properly belong. In acknowledging this he discloses very explicitly his understanding of how this fourth chapter relates to the nine chapters following.[15]

The summary begins with a concatenation of conditionals, of 'ifs' and 'thens', showing that the conditions causally necessary and sufficient for the existence of natural selection are there in the wild, and showing what natural selection tends to bring about over generations. If animals and plants vary, and if there is a struggle for existence owing to Malthusian excess fertility, then, given the complex interactions of species with each other and their physical environments, some variations useful in the struggle will occasionally occur just as variations useful to man occur, and if these variations are advantageous to any organisms in the wild, then those individuals will have enhanced chances of surviving in the struggle and will pass on those advantages to their offspring. This, he reaffirms, is the principle of preservation that he has called natural selection. This evidencing of the satisfying of these conditionals introduces no new causes or effects, but does motivate the shift from settling the existence and adequacy cases to raising the responsibility issue. For Darwin now moves explicitly from conditional to categorical issues in declaring that whether 'natural selection has really thus acted in nature', in modifying and

[15] Ibid., pp. 126–130.

adapting the diverse forms of life, 'must be judged by the general tenour and balance of evidence' in the chapters that follow.[16]

This declaration would have been the right place to end this summary of chapter IV. Darwin has just made the case for those conditions being met in the wild, in completing the cases for the existence of natural selection and for its ability to adaptively diversify species, and in preparing for the case he will make, in chapters IX–XIII, for its having been actually responsible for the production of extant and extinct species. But he cannot resist briefly anticipating these later tasks here, noting that he has already shown how natural selection entails extinction, which geology confirms has occurred throughout the earth's past; and he then goes on to two and a half pages on the explanatory virtues of natural selection, when joined with the principles of divergence and extinction, especially in explaining why animals and plants are classifiable in taxonomic groups with successively subordinate subgroups, a 'great fact' given 'no explanation' on the view that each species has been independently created.[17]

Although deviating fleetingly here from the compositional vera causa ideal, Darwin will knowingly correct this deviation in IX and XIII. When revisiting those chapters near the end of this chapter of our book, we will not charge him with compositional inconsistency, but will be recognising his own recognition of his eventual conforming of his long argument to this ideal.

The Causal–Explanatory Complex Defended and Deployed

Our expectations for the nine remaining chapters of the long argument can now be more explicit; for we expect no new reasoning for and to the selection analogy itself, only further and new reasoning from it. The reasoning to and for the analogy was completed in the first four chapters with the marshalling of the causal–explanatory complex, illustrated and articulated by the diagram and accompanying commentary that ended chapter IV. As this complex persists unrevised for the rest of the long argument, so does the analogy itself.

Chapter V contributes to chapter IV by supplementing the account, in chapters I and II, of the universal laws of variation within and between species, under domestication and in the wild. Opening his summary of chapter V, Darwin famously professes profound ignorance of these laws, admitting that very rarely is it known why a part in an offspring differs as it

---

[16] Ibid., p. 127.   [17] Ibid., p. 129.

does from this part in the parents. But his summary ends by insisting that, whatever the cause may be, one must exist, and that the steady accumulation by natural selection of such differences, when beneficial, is what produces the most important modifications of structure. Chapter V has contributed to the following eight chapters only by contributing to the causal–explanatory complex marshalled in chapter IV. This fifth chapter makes no additions, amendments or amplifications to the selection analogy and its place in that complex; just as chapters VI–VIII make none in defending that complex from objections to its causal adequacy.

Chapters IX–XIII do see amplifications, if not additions and amendments. To see why this might be expected, we may distinguish causal, productive adequacy from explanatory virtue. Establishing that rivers are adequate causes of canyons requires establishing that rivers can produce the canyon-defining features. But, further, one may ask if this river theory of canyon production also has the merit of explaining various further factual generalisations about canyons: that, for example, those nearer the sources of rivers are narrower and shallower than those further downstream.

For Darwin to establish that branching natural selection is an adequate cause of species required arguing that it can cause varieties to diverge sufficiently, adaptively and hereditarily, to count as species. But this theory of species production can be shown to have further evidential virtues if it explains why, say, species peculiar to arid oceanic islands often do not resemble species on arid islands in other distant oceans, but do resemble species on the nearest continental land even when this is not arid, so making it probable that these island species are descended from mainland ancestors, and have actually been produced by natural selection.

Staying with this distinction between the possible and the probable effects of natural selection, recall what issues chapters V–VIII engage. These chapters assume that, in chapter IV, with the aid of the selection analogy, natural selection has been shown to be adequate to change varieties into species, shown therefore to be a possible cause of varieties that count as species. These chapters go on to anticipate and counter objections that species have features, complex adaptive structures for example, such as the eye, that natural selection cannot cause; and that these are difficulties for any case for natural selection being a possible cause of species. Such difficulties are duly countered by arguing that they are overcome when all relevant possible causal consequences of natural selection are taken into account. The case for the adequacy of natural selection as a possible cause of species is thereby vindicated. However, there is no exhibiting of explanatory virtues as evidence for the actual responsibility of natural selection in forming extinct and extant species, evidence such as only the later chapters will provide.

Another distinction helps here. Natural selection has causal consequences; and the theory of natural selection – the theory that natural selection has been the main cause responsible for animal and plant adaptive diversification over eons of past time – has logical consequences. Artificial selection can teach us what causal consequences natural selection has, while the theory that natural selection caused the Galapagos bird species to diverge from their mainland ancestors has, among its logical consequences, explanatory implications which, when confirmed, support that theory evidentially. What explanatory consequences a theory can have depends on what causal consequences the causation it invokes can have. But elaborating and exhibiting the explanatory implications requires argumentation beyond what is needed in establishing those causal consequences. That is why the argumentation in Darwin's chapters IX–XIII goes beyond that in chapters VI–VIII. Those three chapters were arguing in defence of the adequacy case made in chapter IV, so preparing for the responsibility case made in IX–XIII; which is appropriate because natural selection, the theory, cannot explain any generalisations about species unless natural selection, the cause, can produce species.

Natural selection is routinely called by Darwin a cause, but also a law and a principle; and he talks sometimes of a law or a principle being active in producing certain effects, rather than talking of a cause doing so in acting in accord with some law or principle. Any exegesis of the long argument has to give priority to construing the term natural selection as the name of a cause. Laws and principles may have propositional form and content: as with the law of gravitation or the principle of least action in mechanics. Conspicuously there is for Darwin no one law of natural selection, a law which is to this cause as the inverse square law is to the Newtonian gravitational force; although natural selection is not capricious or miraculous, but lawful in reliably causing similar effects in similar circumstances. For Darwin's purposes, conforming his argumentation to the vera causa ideal and the ideal of causal analogy as proportion, natural selection is a lawful causal relation, and a lawful agency or process. In due course, he will say that he uses the term 'much as a geologist does the word denudation – for an agent, expressing the result of several combined actions'. The term has a place in Darwin's and so our language, but, not having propositional form or content, natural selection itself, what the term names, belongs in the world.[18]

In chapters IX–XII, Darwin needs to argue defensively against further anticipated objections to the responsibility case made in these four chapters, because the theory that natural selection has been responsible for species formations cannot be made probable by any general facts about species that it explains, if other facts are inconsistent with its responsibility. He has, in chapter IX, to counter the objection that widespread saltationary discontinuities in the fossil record are inconsistent with any case for new species having arisen slowly from old ones, in gradual descents with continuous modifications by means of natural selection. And, in chapter XI, he has to counter the most manifest and general difficulty posed for the causal–explanatory complex by the geographical distribution of genera, families and wider groups of species: namely, that each such group of related species descends from a single ancestral species in one original location; but that species of the group are now found living in many very distant locations. Darwin has then to argue that migrational dispersals from that one ancestral site could have been extensive enough to result in the present distributions of these descendent species. His arguments appeal partly to dispersal facilitations: birds carrying plant seeds on their feet or in their stomachs, rafts of reeds bearing small mammals across oceans; and partly to past dispersal opportunities, when some mountain ranges were less massive and impassable, or some regional climates were less severe.[19]

With such defensive tasks undertaken, Darwin can move to his inventory of explanatory successes. His summary of these in chapter X fills two pages on how many leading facts of palaeontology follow from the theory of descent with natural selection. He opens with the generalisations that new species appear not in big, sudden batches but slowly and successively; that species of different classes do not always change together, and that extinctions of old species are consequences of the production of new ones, and that when a species has disappeared it never reappears. Likewise in his summary of chapter XII, he emphasises that his theory can explain why in different latitudes, in South America for example, the species living in many very different habitats are nonetheless alike, and resembling also the extinct species found as fossils in that region. Most triumphantly of all he ends his summary of this chapter, the final one of the quartet devoted to palaeontology and biogeography, by arguing that his theory explains the parallelism between regularities in the distribution of species in space and

[18] Hodge (1992a), p. 216.   [19] Darwin (1859), pp. 356–365.

in their succession in time. The more nearly any two forms are 'related in blood' the closer they will be in time and space. In both, the laws of variation were the same, with modifications accumulated by the same cause, natural selection.[20]

Explaining Resemblances and Differences

Like its antecedents, the *Sketch* of 1842 and *Essay* of 1844, Darwin's *Origin* is obviously preoccupied throughout with the resemblances and differences among wild animals and plants, ascribing resemblances often to common ancestry, and differences often to divergent descent and so to natural selection. The book would not have been complete without applying his causal–explanatory complex to these issues as treated by taxonomists in their classifications, by morphologists in their comparative anatomies, by embryologists in their generalisations about individual developments, and by naturalists in their studies of rudimentary or vestigial organs. All this he does in chapter XIII.

In two chapter summaries Darwin had briefly prepared readers for the views he would take in chapter XIII. Closing chapter VI, the first on the difficulties facing his theory, he announced that all organisms have been formed on two great laws: Unity of Type and the Conditions of Existence. He associates Georges Cuvier with the second, and, as he will later, he should have associated Cuvier's opponent Etienne Geoffroy Saint-Hilaire with the first. Unity of Type is, he says, the fundamental agreement in structure in organisms of the same class, quite independent of their habits of life, this unity being explained, for Darwin, by unity of descent, whereas natural selection adapts species to their conditions of existence. On his view, then, Conditions of Existence is the 'higher law' as it includes, through the inheritance of former ancestral adaptations, Unity of Type.[21]

These brief assertions closing chapter VI need glossing. For millennia various explanations had been offered for organisms' resemblances and differences. From ancient times to Darwin's day, some naturalists had said that animals similar in structure, bears and deers say, were so because they were similar in the degree of their complexity and perfection, and so close in their position in the *scala natura*, while those that were very different, fish and worms, say, were so because they were far apart in the *scala*. Such was the levels view. Cuvier had opposed all such views, urging instead that structural similarities and differences are due to adaptational similarities

[20] Ibid., p. 410.   [21] Ibid., p. 206.

and differences, birds almost all being adapted to flying while fish are to swimming. This is comparative-anatomical adaptationism. Against Cuvier, Geoffroy insisted that similarities are due to sameness of plan or type, fish and birds being structured according to a common plan. The various different plans structuring even the most diverse animals are indeed, he held, variants of a single fundamental plan for all.[22]

Darwin's agreement with Geoffroy about unity of type is therefore limited and overridden by his alignment with Cuvier. For when, as the creationist Cuvier would not have done, ancestry and history are ascribed by Darwin to unity of type as unity of descent, that unity yields priority to adaptation to conditions of existence; since, for Darwin, the structure of a common ancestral species was originally a result of earlier adaptive natural selection. Darwin's integration of common ancestry and divergent selection has debts to the common plan view and to the adaptationist view, but none here at least to any common levels view.

These Darwinian commitments allow us to map his views in relation to historicism, functionalism and structuralism as a trio of alignments found in many fields in his century, including architecture and linguistics as well as biological science. Darwin integrates his functionalism (adaptations and their causes and effects) with his historicism (ancestries and genealogy and their consequences) while making his Geoffroyan structuralist conceptions decisive but not as fundamental as functional and historical considerations.

Darwin had closed chapter IV by integrating classificatory taxonomies and arboriform genealogies for ancestral and descendent species. There, as in the more extensive treatment of classification in chapter XIII, he argued that this integration entails an explanatory and so evidential triumph for his theory. For his theory explains why animals and plants, extant and extinct, can be classified as they can be. Classifying is grouping and subgrouping: the birds are a subgroup of the vertebrates, and within that subgroup, the finches and gulls are again subordinate groups. The narrowest, most subordinate, groupings are specific: the chaffinch or the herring gull species. The widest groupings are kingdoms: animals or plants.

Darwin's integration of classification and genealogy rests on one thesis. The various species in a narrow grouping have descended from a single recent common ancestral species, with little time since for branchings, divergings, multiplyings and extinctions of descendent species; a wider grouping traces to a more remote single common ancestry with many more subsequent multiplicative and divergent descents of species belonging to

---

[22] Appel (1987).

diverse supra-specific groups. His genealogical explanations all invoke this integration. Extinctions are decisive here, as real causal and historic processes and patterns, most consequentially in contributing to the producing of ever larger gaps both within and between those groups that have seen the most extinctions over most eons.[23]

In chapter XIII Darwin argues that a natural classification is one conformed to genealogy, for a natural classification groups according to affinities not analogies. In their internal organisation whales share many affinities, many fundamental mammalian characters, with monkeys, and only a few superficial characters, analogies, with sharks. This affinity–analogy contrast was a recent innovation in Darwin's day and its rationale in systematic, taxonomic natural history had no direct bearing on analogy in the sense of Darwin's selection analogy. In moving from classification to morphology, he works with an even more recent contrast between homologies and analogies, which again had no such direct bearing. A monkey's hand, a whale's flipper and a bat's wing are homologous structures with a common pentadactyl structure. A butterfly's wing is not so structured, and is therefore analogous to the bat's wing because they are only functionally similar.

It is affinities and homologies and not analogies that are due to common ancestries and so are most valued in classifying in accord with genealogy. A natural classification was often thought to be one in accord with the plan of creation; but Darwin does not see genealogy as delineating any plan. There is no regularity in branching descents and divergences, and no tendency, then, for all groups to have a standard number of subgroups, five, say, as in the so-called quinarian schemes; nor will a mapping of groups according to their close or remote affinities conform to any regular figures such as the circles of those schemes.

In his treatment of comparative embryology Darwin again avoids endorsing regularities other naturalists had cherished. Foetal mammals have gill slits before they develop lungs. And Darwin takes this sequence as evidence for a fish ancestry for mammals; but he does not join those who saw the sequence as evidence for a developmental law of form conformed to by individual mammal development today, and by mammal species succeeding fish species in the long ages of terrestrial history. Darwin's not embracing this doctrine is required by his commitment to branching natural selection. Only one line of fish diversification included any species ancestral to the mammals; all the other lines did not. The

---

[23] Darwin (1859), pp. 128–130 and 411–434; Winsor (2013).

genealogical succession from fish to mammals is exceptional and not typical, and not conforming therefore to any developmental law of form. It was natural selection that caused some exceptional fish species to have their atypical mammal descendants.

Is there, then, any lawful tendency common to individual mammal developments today and these mammals' past descent from fish ancestors? For Darwin the law is divergent specialisation. Von Baer had insisted on distinguishing the level and the characters of any animal organisation, and had emphasised that animals with the same level or degree of functional and structural differentiation could be very unlike in the character of their organisation. In line with this distinction, Darwin could hold that individual mammal developments today, and the ancestral mammalian diversification over past eons, could both show a succession from less to more functional and structural specialisation. And if such specialisation is the criterion of progress then this is a tendency to progress in both – as biologists will soon be saying – ontogenies and phylogenies, but with no assumption that progress for all fish species is always destined to include mammalian descendants. All mammals have a common fish ancestry in their past, but only a very few fishes have had mammalian descendants in their future. Ontogenetic developments may then reliably and lawfully recapitulate eons of past descent, without this descent by means of natural selection being lawfully and reliably repeatable phylogenetically.[24]

For this integration of common descent and branching natural selection, the most telling embryological generalisation is that embryos resemble one another more than do the mature individuals that they become. This greater resemblance is true within any species and between one species and another. And this is especially true in domestic animals. Embryonic dogs of all varieties are much more alike than the terrier and greyhound adults they grow into later. And prenatal puppies and kittens are more alike than are adult dogs and cats.

In articulating his final explanatory-evidential triumph, Darwin ascribes these embryonic resemblances to community of descent, and the adult differences to diversifying selection. And likewise with wild animals: natural selection diversifies adults more than embryos. Especially is this true of placental mammals whose embryos all live in very similar uterine conditions, while postnatals differ in the various ways of life and circumstances to which natural selection has adapted them. Natural selection has been able to produce these adult differences because not all variations originally occur early in life, and all are inherited so as to recur subsequently at the same earlier or later prenatal or postnatal times. In lower animal species with free-living larval stages natural selection has adaptively diversified those larvae in fitting them to different environmental conditions.[25]

This synthesis of ontogeny and phylogeny through the integration of common ancestry and diversifying selection draws on passages in chapter IV, where Darwin compares the relations between variation and selection in artificial, in natural and in sexual selection. So it makes a fitting final triumph for the concluding of the book's one long argument, given how that argument has pivoted on that later moment at the closing of chapter IV, the moment when the marshalling of the causal–explanatory complex ended and gave way to its defence and deployment, all in accord with the Reidian vera causa evidential ideal and the Aristotelian ideal of causal-analogical proportionality.

Chapter XIII and the long argument begin their ending with five pages on what Darwin sometimes calls rudimentary and sometimes atrophied or aborted organs. He does not have the later distinction between rudimentary and vestigial organs, and calls rudimentary the mammae on many male mammals, the teeth in foetal whales absent in adults, the tiny useless wings in some insects and any milkless teats on domestic cows. His explanations for these and many other examples get only a few sentences in the chapter summary, because these explanations have invoked no special principles beyond those expounded in his earlier embryological reasoning. Larvae, he reminds us, are active embryos specially modified in relation to their habits of life, in accord with the principle that hereditary modifications are inherited at corresponding stages. Given this same principle; and given that organs reduced in size, either from disuse or selection, will be reduced at the time in life when the organism is providing for its own wants; and given how strong is the principle of inheritance, then rudimentary organs and their final abortion present no inexplicable difficulties; and the importance in classification of these organs, and of embryonic characters generally, is intelligible on the view that a taxonomic arrangement is only as natural as it is genealogical.

This reflection leads neatly to a resounding closing declaration on behalf of the long argument: the several classes of facts considered in this chapter seem to Darwin to proclaim, so plainly, that all the world's species, genera and families have all descended from common parents each within its own

[25] Ibid., pp. 446–450.

class or group, that he would 'without hesitation adopt this view, even if it were unsupported by other facts or arguments.'[26]

The recapitulation in the first half of the last, fourteenth, chapter is helpful in surveying the main facts supporting the theory of natural selection and those presenting difficulties, and in giving Darwin a chance to say why he sees the positive considerations outweighing the difficulties when these are overcome in his countering of them. But his survey is unhelpful in that he makes no allusion to the two part and three case structure of his vera causa argumentation; so the reader gets no aid here in appreciating how that structure has informed the sequence of his book's chapters.

Nor does the second part of the final chapter include such elucidations in its instructive responses to three issues only taken up in the closing pages of the book: the reasons why most naturalists have rejected the transmutation of species; his own reasons for extending common descent to include all plants and animals in one tree of descent from a single, original, first species; and his predictions for the sciences of life if, or rather when, his views eventually become widely accepted, especially by young men more open to novel notions than their older mentors.

In the opening paragraphs of the *Origin* his rhetoric was predominantly the rhetoric of *ethos*, of character, his character as an authoritative, gentlemanly naturalist and traveller, diligent and disciplined as a man of extensive factual experience, and cautious and patient as a contemplative theorist. The rhetoric of the book's closing paragraphs is often the rhetoric of *pathos*, of feelings. If viewed not as special creations but as descendants of more ancient ancestors, animals and plants, past and present, become ennobled. There is grandeur in this, his view of life. It is tempting then to suggest that the thirteen chapters of the long intervening argument complete the classical trio of persuasive arts by presenting the *logos*, the reasoning, as the beef in this burger. But no comprehensive account of its rhetoric would sustain this trivialising of the book's compositional strategies, as confirmed implicitly in our next two chapters.[27]

Here we may take a last glance at the divisions within the long argument. As just recalled, the main one is between the marshalling of the causal–explanatory complex completed in chapter IV, and the defending and deploying of that complex in the next nine chapters. In the first four chapters, as supplemented by chapter V, the evidential cases are made

---

[26] Ibid., pp. 457–458.
[27] On Darwin's rhetoric in the *Origin*, see Depew (2009); Fahnestock (1996).

for the existence of natural selection as a vera causa, and for its adequacy, its competence, as a cause of new species. The completed complex integrates common descent with natural selection as the cause of branching divergences and terminal extinctions. The adequacy thesis is defended against difficulties in chapters VI–VIII. The explanatory evidence, the evidence from the explanatory virtue of the causal–explanatory complex, and so for the responsibility of natural selection in producing new species in remote and recent times, is presented in chapters IX–XIII. In chapter IV the selection analogy supports the adequacy case directly; and then later contributes to the responsibility case indirectly, by evidencing not just what natural selection the cause can produce, but what natural selection, the theory, can explain.[28]

This sequence of divisions within Darwin's long argument shows why the book is ordered as it is. As with Newton's *Principia* and Lyell's *Principles* there is one major division of expository labour: all the causal–explanatory resources are first marshalled and then, second, they are defended and deployed. Darwin's marshalling proceeds from existence questions to adequacy ones, in that order because only existent causes can be adequate causes. His defence and deployment tasks proceed as they do because if objections to the adequacy claim are not countered, explanatory deployment cannot go on to argue from explanatory success to causal responsibility. For confirmation that Darwin saw this cumulative structuring as the only way for the book to be written, one may read its chapters in reverse order, from XIV to I. Read that way the chapters simply do not add up to cumulative argumentation conforming to the vera causa evidential and expositional ideals.

---

[28] For incisive further discussion of the ordering of the *Origin's* exposition, see Sober (2010).

CHAPTER 5

An Analysis of Darwin's Argument by Analogy

In Chapter 4 we saw that the *Origin* has an overall structure conforming to the vera causa evidential and compositional ideals. First, the existence of natural selection is established, and then its power to make new varieties and species. Finally, there are arguments and evidence to confirm that it has been responsible for making both older extinct and more recent extant species. Here, in this chapter of our book, the focus is on the second part of this structure, and so on the argument, the reasoning, establishing that natural selection has the power to make new varieties and species. This is argument by analogy with artificial selection, which is known to be an adequate cause to have the power to make new varieties. This invocation of analogy is, we will be claiming, an *argument*, and not merely rhetorical or heuristic in its function. And it is an argument by analogy in accord with the classical Greek conception of analogy.

We will, then, be analysing how the argument is designed to give direct empirical support to the theory of evolution by natural selection. This evolution is a process taking millions of past years at a rate far too slow to be observable, whereas artificial selection is a familiar fact, empirically observable and authoritatively investigated. If artificial selection can be shown to produce fresh varieties, then so can natural selection, provided that artificial and natural selection can be established as analogous in relevant respects.

Suspicions about Analogical Argument

> Analogy would lead me one step further ... But analogy may be a deceitful guide.[1]

Argument by analogy has a bad reputation. Everyone will agree that analogy can have a valuable role in science in suggesting profitable lines

---

[1] Darwin (1859), p. 484.

of enquiry and perhaps a pedagogic role in explaining scientific theories to the public. There is, however, a widespread suspicion of *argument* by analogy, and of granting it any evidential or probative status in the establishment of a scientific theory.

There are a number of reasons for this suspicion, some good, some bad. The worst reason, which we believe may still persist, is a hangover from the Enlightenment reaction against mediaeval scholasticism. There was then a wish to replace the mediaeval methods of enquiry, in which analogy had played a prominent role, with good inductive science. Here as elsewhere, whatever advance in understanding the Enlightenment represented, there was also a loss of insight.

It was in the wake of the Enlightenment abandonment of scholastic theories of analogy that the kind of account of argument by analogy that was given canonical representation by Thomas Reid emerged. The idea that *this* is what an argument by analogy is widely prevalent even today. For instance in the *Stanford Encyclopaedia of Philosophy*, Paul Bartha writes:

> An *analogy* is a comparison between two objects, or systems of objects, that highlights respects in which they are thought to be similar. *Analogical reasoning* is any type of thinking that relies upon an analogy. An *analogical argument* is an explicit representation of a form of analogical reasoning that cites accepted similarities between two systems to support the conclusion that some further similarity exists.[2]

This is the only account of 'analogical argument' that Bartha will offer and go on to examine. The original conception of analogy as conforming to the formula 'A is to B as C is to D' is not even mentioned in the entire course of the article. Once Bartha's account is accepted as what argument by analogy is, suspicion of such arguments, or at least extreme caution in their use, is fully justified. Such arguments are at best suggestive, indicating possible directions for future investigation. Not only are they weak arguments, they are wholly inadequate for giving a proper account of the argument we are concerned with in the *Origin*. One reason for the unclarity surrounding the question of the precise nature of Darwin's argument is that many of the commentators have assumed that when we discuss whether the argument we find in the first four chapters of the *Origin* is an argument by analogy, what we are talking about is an argument of the form we have just looked at, or something like it. In what follows we shall set this form of argument on one side, and when we

---

[2] Bartha (2019).

talk of an argument by analogy we shall always mean the form of argument analysed, for example, by Whately in the early nineteenth century. Here, as on other topics, he was deliberately following Aristotelian precedents.

However, even when we look at argument by analogy in the strict sense, it is necessary to be cautious. Even in this form, there are many examples of faulty arguments and an author such as Whately, alongside his analysis of examples of such arguments that are sound, will devote even more space to the abuse, or misuse, of arguments by analogy. We shall see more clearly what is at stake here if, before turning to Darwin's argument, we look at the ways in which such arguments can go wrong.

A schematic version of the form of argument by analogy that concerns us is as follows:

> Situation M is an analogical model of situation T.
>
> M is F.
>
> *Being F* is invariant under analogy (may be transferred from M to T),
>
> ∴ T is F.

Here we have an argument form that is trivially valid. If such an argument goes wrong, it can only be because no proper justification is given for one or more of the premises. Either there is no adequate reason for accepting we are here dealing with an *analogical* model, or no justification is given for thinking that the property of *being F* is one that can be transferred from the model to its target.[3] Let us look at examples of each kind of failure.

First, we must ensure that we really have an analogical model. Consider the following metaphor, apparently based on analogy, as an example:

> Memory is a net: one finds it full of fish when he takes it from the brook, but a dozen miles of water have run through it without sticking.[4]

Here we clearly have a superficial structural correspondence between two different situations, so that we can regard the brook as giving us a model for experience. But the question is, 'Is it an analogical model?' That is to

---

[3] For what follows, compare Joseph (1916), p. 532: 'If the relation really is the same in either case, then what follows from the relation in one case follows it in the other; provided that it really follows from the relation and nothing else.' Joseph's whole treatment of argument by analogy (pp. 532–542) is still one of the best known to us. Also, despite considerable conceptual unclarity in the earlier parts of the article, Brown (1989), pp. 169–171 has a highly intelligent discussion of an argument that is worth considering in this context: 'An innkeeper is to a guest as a steamboat proprietor is to a passenger; an innkeeper is strictly liable for any theft or fraud suffered by a guest; therefore, a steamboat proprietor is strictly liable for any theft or fraud suffered by a passenger.'

[4] Oliver Wendell Holmes, jr.

say, can we regard the formula 'memory is to experience as a net is to a brook' as a genuine analogy? To rephrase that, is there a relation R such that R(memory, experience) = R(net, brook)? What is an appropriate value for 'R'? It seems impossible to give a sensible answer to that question (or rather, it seems possible to give two, incompatible answers. For the metaphor to be coherent we should require that the relation of a net to fish should be the same as the relation of a net to water, which is clearly not the case.) But without that, it is impossible to construct any argument by analogy telling us anything further about the relation of memory to experience than was already encapsulated in the superficial structural correspondence between the two situations: we only remember a small fraction of what we have experienced.

However, far and away the most frequent failure of an apparent argument by analogy is a failure to justify the third premise in the above schema: '*Being F* is invariant under analogy (may be transferred from A to B)'.[5] Without a justification for this premise, the argument simply collapses into a variant of Reid's argument. In the mathematical case that we looked at in Euclid Book VI, this premise can be established by a geometrical theorem, but outside that context, we can have no a priori guarantee of its truth. It was for precisely that reason that Kant stressed the difference between the mathematical and non-mathematical uses of analogy, and expressed scepticism about the possibility of a sound argument by analogy outside mathematics. If that was Kant's position, however, he was being overhasty. What is required here is what Mill called 'the *material* circumstance'[6] – 'to be that on which all the consequences, necessary to be taken into account in the particular discussion, depend'. The difference between the mathematical and the non-mathematical case is simply that in the non-mathematical case this requirement concerning the material circumstance cannot be established a priori, but requires further empirical support.

---

[5] A clear example illustrating the contrast between an analogical model that permits a valid inference from model to target and one that does not is provided by Gentner and Gentner (1983). Two groups of subjects were shown an electrical circuit and asked to work out the flow of current. Both groups were told the relevant electrical theory (Ohm's Law, etc.) and also given an analogical model for the circuit. For the first group the analogy was with water draining through pipes from a reservoir, but for the second with people running along corridors. Members of the first group arrived at the correct answer to the question, but members of the second failed. The explanation for this is clear, in the case of water pipes gravity provides an analogy for voltage in the electrical case, whereas we are told nothing about what drives the people in a way that would correspond with voltage, making any inference impossible.

[6] See Mill (1843) Book III, ch. XX quoted at the end of Chapter 2.

Mill, like Whately before him, illustrates this idea with plausible examples. We shall attempt to give a more explicit account of the form that an argument by analogy now takes in the light of this. What is required is to explicitly include a reference to the material circumstance in the statement of the form of the argument. The material circumstance will be something true about M that makes it true that M is F, but which is also true of T. Let us suppose that this is the fact that M is G. (It is precisely because M is G that it is F, and T is G). We may now expand our argument as follows:

Situation M is an analogical model of situation T.

M is F.

M is G.

The fact that M is G is the sufficient cause or reason for M's being F.

T is G.

∴ T is F.

Many proverbs provide simple examples of such uses of analogy. Consider, for example, 'A stitch in time saves nine'. This proverb is freely, and validly, applied by analogy to the most diverse situations – wherever there is a problem that, left untreated, will get out of control. Thus, if rising life expectancy makes pensions unaffordable, retirement ages must be raised; and if this is not done early enough, the problem escalates. Here 'F' is 'a problem that must be dealt with early if it is not to get out of control', 'G' is 'a problem that gets worse if untreated', and the basis for the first situation being an analogical model for the second, is that 'darning is to a hole in a garment as raising retirement age is to the affordability of pensions'.

We may see further how this works out in practice by looking at a case of an argument that is defective just because the 'material circumstance' is not shared by the model and its target. We shall consider an argument that Francis Galton set great store by in support of a 'saltationist' conception of the evolutionary process in which, by contrast with Darwin's insistence that evolution was a slow and gradual process, it was possible to have sudden transitions from one state of evolutionary equilibrium to another different state, and that you could have new species arising as a result of an event that causes a violent shock to the system:

> The mechanical conception would be that of a rough stone, having, in consequence of its roughness, a vast number of natural facets, on any one of

which it might rest in 'stable' equilibrium. That is to say, when pushed it would somewhat yield, when pushed much harder it would again yield, but in a less degree; in either case, on the pressure being withdrawn it would fall back in to its first position. But, if by a powerful effort the stone is compelled to overpass the limits on the facet on which it has hitherto found rest, it will tumble over into a new position of stability, whence just the same proceedings must be gone through as before, before it can be dislodged and rolled another step onwards. The various positions of stable equilibrium may be looked upon as so many typical attitudes of the stone, the type being more durable as the limits of its stability are wider. We also see clearly that there is no violation of the law of continuity in the movements of the stone, though it can only repose in certain widely separated positions.[7]

We may here slightly simplify Galton's analogy, without affecting his argument. We take the case of a die, which has six different states of stable equilibrium, depending on which of its faces is uppermost. If we give the die a small tap, then it will rock but return to the same state. If, however, we give the die a hard knock, it will topple over, landing with a different face uppermost. Here we see how something can be moved from one state of stable equilibrium to another as a result of a single, sufficiently large, shock. By analogy, it is now made comprehensible that a population should move from one state of evolutionary stable equilibrium to another as a result of a single, sufficiently large, shock.

It is possible to quarrel with this argument by asking whether we really mean the same by 'a state of stable equilibrium' in the two cases. But we can leave that question on one side, since there is another more decisive flaw in this argument. When we ask 'What is the "material circumstance" in this argument?', we are asking what makes it true that it is readily possible to switch the die from one state of equilibrium to another by applying a single sufficiently large shock to the die. The answer is that it is precisely because the die has a set of already established states of equilibrium that are already features of the die before it receives the shock. It is because of this that it is simply a matter of a single switch to move from one state to another. However, this is not a feature possessed by the population that is to evolve. In a transition from one state of evolutionary equilibrium to another, a population has to create the second state ab initio and that is prima facie a far more complex process than a die turning over.[8]

[7] Galton (1869), p. 369.   [8] Cf. Dobzhansky (1937), p. 40.

In the light of this, whatever else can be said for or against saltationist conceptions of the evolutionary process, this argument by analogy is worthless. We must now see how Darwin avoids the pitfalls surrounding the use of argument by analogy.

The Analogy between Artificial and Natural Selection

Although Darwin himself runs these stages together in his exposition, it is convenient and makes for greater clarity to break his argument down into two stages. The first stage is to show that the argument by analogy establishes that just as artificial selection (henceforth 'AS') creates new varieties of plants and animals, so natural selection (henceforth 'NS') can create new varieties. The second stage, building on the first, is to establish the stronger claim that unlike AS, it is possible that NS should be able to create new *species*.

We first establish the analogy between AS and NS, between the way animals and plants are treated under domestication and the way they fare in the struggle for existence in the wild.

Domestic Breeding

The facts that Darwin draws upon for his account of domestic breeding, or AS, are all either familiar everyday facts, or what he has learnt from extensive discussions with those who engage in such breeding, such as pigeon fanciers. What we have to do is assemble a set of such facts as are relevant to the analogy that Darwin is to make.

The most fundamental fact for the whole theory that Darwin is developing is the fact of *variation* – the fact that offspring are imperfect replicas of their parents. It is a familiar enough fact in the domestic setting that offspring possess traits not present in their parents, or possess those traits to a different degree. Unless such variations existed, change would stop. Equally important for the theory is that a large number of such variations may be *inherited*: if John's son is taller than John then that increases the probability that John's grandson would also be taller than John. If this were not so, then any change would be stopped in its tracks after only one generation.

The next basic point is that some of these inherited traits are desirable, some undesirable and some indifferent. Here what counts as a desirable trait is always from a human point of view: what is a desirable trait in an animal depends upon the use to which the animal is to be put. What that means is that there are traits that one human will treat as desirable, but another undesirable: very different qualities are needed to make a good guard dog, sheepdog, guide dog, retriever or pet.

Against this background, we next introduce the fact that human beings *select*. In this context this means that in a variety of ways human beings discriminate among the plants and animals that they breed in such a way as to increase the probability that those with desirable characteristics will survive and reproduce, and decrease the probability that those with undesirable characteristics do so. We may first consider what Darwin calls 'conscious' selection: the practices of those who are deliberately attempting by their selections to produce new varieties of the creatures they tend: it is both the case where it is completely straightforward to give a description of what happens and most importantly the case for which Darwin would be able to obtain direct empirical evidence. We may think here of a pigeon fancier wishing to produce a long-beaked variety of pigeon. They will select the pigeons with the longest beaks available to breed from, and then repeat this consistently and persistently through a series of generations. The result is that the pigeons' beaks will tend to get longer and longer, until the beaks are long enough that the result can count as a new variety of pigeon.

Alongside this deliberate selection in order to produce new varieties, Darwin introduces another kind of selection, which he calls 'unconscious selection'.

> At the present time, eminent breeders try by methodical selection, with a distinct object in view, to make a new strain or sub-breed, superior to anything existing in the country. But, for our purpose, a kind of Selection, which may be called Unconscious, and which results from every one trying to possess and breed from the best individual animals, is more important.[9]

Throughout human history, human beings have in a variety of different ways given preferential treatment to those animals among their stock that best suit their purposes – the use they wish to make of them. This, Darwin claims, has led to the production of the different varieties of those animals that we see around us today. The first point to make about unconscious selection is that whereas we are in a position to observe methodical selection and its result directly, the evidence for the claim that unconscious selection has been responsible for the present array of varieties of domestic animals and plants is necessarily indirect. All that we can actually observe is

---

[9] Darwin (1859), p. 34.

that present array.[10] What we find for each domestic species is a number of distinct varieties, each differing markedly from one another and from the wild members of that species. Each of these varieties is, however, well-tailored to be put to a specific use. The only possible explanation for this array is that it is a product of human beings having throughout history continually selected for animals and plants best suited for their purposes, without necessarily being conscious that they were thereby generating fresh varieties.

But if the evidence for unconscious selection is necessarily indirect, why not restrict attention to the case that can be observed directly, conscious selection? A variety of answers are possible here, such as that the production of new varieties by unconscious selection shows that new varieties can be generated by a process in which the intention to produce them plays no part. However, one crucial reason for not doing so can be found in the following passage:

> In man's methodical selection, a breeder selects for some definite object, and free intercrossing will wholly stop his work. But when many men, without intending to alter the breed, have a nearly common standard of perfection, and all try to get and breed from the best animals, much improvement and modification surely but slowly follow from this unconscious process of selection, notwithstanding a large amount of crossing with inferior animals. Thus it will be in nature.[11]

If we restricted our account to conscious selection, we would have no answer to the challenge, 'Selection creates new varieties in the unnatural conditions adopted by the pigeon fancier, but this only happens because the pigeon fancier carefully controls the process in such a way as to prevent any crossing of the desirable pigeons with inferior stock. However, in the natural world there will necessarily be much crossing of creatures with desirable traits and those without those traits. How do we know that that will not counteract the effect of NS?' If, however, unconscious selection does slowly but surely over the long course of human history succeed in producing the varieties with which we are familiar, then, since that history will be replete with cases of crossing desirable animals with less desirable animals, such crossing cannot be a barrier to the ultimate success of selection in the rough and tumble of the natural world.

There is another feature of AS that stands out more clearly in the case of unconscious selection than in that of methodical selection. When we look

[10] Ibid., p. 40: 'we know nothing about the origin or history of any of our domestic breeds'.
[11] Darwin (1859), p. 102.

at the differences between a carthorse, a racehorse and a wild horse, say, we find that they do not simply differ in a single trait, but rather there are a whole range of differences, a whole range of traits that are interrelated in complex ways so as to produce an animal that is as strong as a carthorse or as fast as a racehorse. We therefore must reckon with a complex process of selection, where several traits are selected for simultaneously so as to produce the animal that functions in the way with which we are familiar.

To summarise: we may contrast methodical and unconscious selection along the following lines. Methodical selection is directly observable, produces dramatic changes quickly, small scale, but needs careful control to prevent intercrossings that can vitiate its effect. Unconscious selection is indirectly inferred as the best explanation of the varieties we see around us, works slowly, being spread across human history from its beginnings to the present day, is large scale, but without steps being taken to prevent intercrossing. There is no *programme* of unconscious selection: it is instead constituted by the whole amorphous history of human dealings with plants and animals, a widespread tendency on the part of human beings, whenever they have a choice, to choose the animals best suited to their purposes. The importance of methodical selection for the argument lies in the fact that it is directly empirically available, and can be observed and experimented with. The importance here of unconscious selection is that, being uncontrolled, it provides the perfect analogical model for NS. If the argument relied exclusively on methodical selection, it would be vulnerable to the criticism, that it only showed that a programme of selection can produce new varieties under carefully controlled circumstances, where there is no such control in the wild, where NS is supposed to operate.

The Struggle for Existence

Darwin must now show that an analogically comparable set of facts obtain in nature to form the basis for setting up an analogical model between what happens to domestic animals and what happens in nature. Specifically, he must show that there is variation in nature, and that there is a process parallel to AS.

Darwin did not claim to know why particular individual animals differ as they do from their parents and from each other. But, as shown in Chapter 4, he held that all hereditary variation on the farm or in the woods is due to variation in environmental conditions: soil, climate, vegetation, food and so on. And he conjectured that these factors acted through nutritional influences on the male and female reproductive elements prior to fertilisation and conception. These causes were most effective and their effects most observable in domestic animals. But Darwin appealed to Lyell's geology as establishing that changes in environmental conditions are everywhere and always taking place, and so must be causing hereditary variation in wild animals as they do in domestic ones, variation that is cumulatively selectable whether by the struggle for existence or by a stockbreeder.

The main point is, however, to establish the existence of 'NS', a process parallel to AS. It is here that the ideas of Thomas Malthus were crucial for Darwin. 'It is the doctrine of Malthus applied with manifold force to the whole animal and vegetable kingdoms; for in this case there can be no artificial increase of food, and no prudential restraint from marriage.'[12] 'The doctrine of Malthus' that Darwin refers to takes its starting point in a mathematical proof that a geometrical progression will always eventually outstrip an arithmetical progression.[13] Thus the interest on a modest loan with a low level of compound interest will always *eventually* cost more than the interest on a large loan with an exorbitant level of simple interest. This is then applied by Malthus to the situation of a human population. Suppose human beings in a certain environment reproduce at a higher than replacement rate. Then even if initially the environment is rich enough in resources to sustain the entire population, eventually, since the size of the population will increase geometrically but the amount of resources to sustain that population will only increase arithmetically, there will be more people than can survive in that environment, leading to famine or some other catastrophe.

Both Darwin and Malthus are assuming that left unconstrained there will be a much higher than replacement level of reproduction both among humans and animals. In the human case, the most disputable premise is that 'the amount of resources to sustain that population will only increase arithmetically'. However, as Darwin observes, animals and plants are incapable of taking steps to increase the resources available to them, so that in this case the doctrine applies 'with manifold force'.

[12] Ibid., p. 63. See Malthus (1798).
[13] That is to say, suppose we have two populations, the first with initial size A, and an incremental increase in size through successive generations x, and a second population with initial size B, and an incremental increase in size through successive generations y. But where the first population increases geometrically, and the second arithmetically, then the first will always eventually outstrip the second in size. (In the long run even modest compound interest will cost more than exorbitant simple interest.)

If a certain population of animals lived in an environment that was such that the entire population could live out their lives and reproduce without difficulty, then there would be no pressure on that population to change their characteristics, and each generation would possess precisely the same set of characteristics as the previous one, so that in these circumstances there would be no form of evolutionary change. The animals would simply continue in the same state in perpetuity.

However, applying Malthusian considerations shows that such a happy state could not persist and animals would be driven out of Eden. The situation would inevitably arise in which there were more animals in the population than there were resources that could sustain the population, and if unchecked, given the explosive power of an exponential increase in the size of the population, eventually there would only be sufficient resources for a tiny fraction of them. Clearly, once the situation has arisen in which resources were too scarce for the whole population, not all of the members of that population could survive. There thus arises a struggle for existence, in which animals and plants have to struggle with the various difficulties that they encounter in their environment. This sets up an indirect competition between the different members of the population, with the winners being those who struggle most successfully.[14] The ones most likely to survive in that situation would be those best equipped to cope with the various obstacles that tend to prevent creatures from living long enough to reproduce or best equipped to exploit the favourable features of the environment. Thus animals that were agile enough would escape predators, drought-resistant plants would fare better in the desert, and plants that could better attract insects to pollinate them would each have an advantage over the rest.

This enables us to see the various obstacles and favourable features of the environment as having a role parallel to selection in the domestic case: just as those animals that the farmer favours are the most likely, as a result of the consequent favourable treatment, to be able to survive and reproduce, so those animals and plants that cope best with the obstacles that confront them are the most likely to be able to survive and reproduce. Thus in the same way that human breeders discriminate in favour of creatures with the traits they deem desirable, the obstacles confronting

[14] To understand what is meant by struggle here, it is important to recognize that the competition is indirect. This is not typically a question of 'nature red in tooth and claw'. The struggle is not in the first instance with other members of a creatures own species, but with the difficulties in the way of existing successfully that the environment presents to them. It is those who are most successful in *that* struggle who will thereby 'defeat' other members of their species.

creatures in nature discriminate in favour of creatures with the traits that best enable them to cope with those obstacles.

Artificial Selection as a Model for Natural Selection

We start out from a range of straightforward analogies between human selective activities and the discriminatory effect of the struggle for existence, along the lines 'This farmer is to these cattle as this desert is to these plants' in that both act in such a way as to favour some animals or plants over others, increasing the probability that such animals or plants will survive and reproduce. However, the process that concerns us – the way that selective breeding produces new varieties – involves not simply single acts of selection, either by humans or by Nature, but a succession of such selective acts all pointing in the same direction and favouring animals or plants possessing the same traits. Hence we need to move beyond the individual analogies to models involving multiple analogies.

The resulting analogical model between breeding under domestication and the struggle for existence in nature has a more complicated structure than the models we encounter within mathematics such as using one triangle as an analogical model of another similar triangle. There we had a straightforward isomorphism between two geometrical figures, correlating three sides of the one triangle with three sides of the other, with analogical relations between corresponding pairs of sides. Here there can be no question of a simple isomorphism. There obviously are or have been vastly more creatures in the history of the world than those that have been domesticated and the evolutionary processes yielding new varieties and species in nature involve vastly more generations of creatures being selected than those involved in the relatively short period of human history. What we have on the farm are a vast array of sequences of selection, in which successive generations of animals and plants possessing traits that the farmer deems desirable are given favourable treatment, improving the prospects that they will produce more offspring than the animals and plants lacking those traits. Within each sequence the trait or traits selected will gradually become more pronounced in successive generations, until eventually the traits in question are sufficiently well marked to constitute a new variety, so that starting originally with wild horses, farmers selecting for strength will breed carthorses, while race owners selecting for speed will breed race horses. When we turn to the world of nature, we can identify a similar vast array of features of the environment that will discriminate in favour of animals or plants possessing certain traits at the expense of creatures lacking those traits – the winter favouring the hardiest plants, the desert the most drought-resistant plants, the best camouflaged mice tending to be those that are missed by owls, and so on. What is more we shall have similar *sequences* of such discriminations, with, say, successive generations of hardy plants surviving successive winters. We can thus use the vast array of sequences on the farm as an analogical model for the network of sequences in nature.

We may now regard the following points as having been established: There are hereditary variations in plants and animals both those kept under domestication and in the wild; human beings in a variety of ways engage in selective practices that discriminate in favour of animals and plants possessing traits that they regard as desirable, making it more likely that these creatures will be able to reproduce than other animals and plants; as a result of these selective practices there eventually arise animals and plants possessing those traits to a sufficiently marked degree as to constitute a new variety of the species to which they belong; and finally we can establish the situation of domestic animals and plants as an analogical model for the situation of animals and plants in their natural state, and what is more a *working* model. What we have now to show is that we are justified in inferring by analogy from the domestic situation to its wild counterpart that just as AS will produce new varieties on the farm, so the NS effected by the struggle for existence will produce new varieties in the wild.

The 'Material Circumstance'

Kenneth Waters, in one of the very few accounts known to us to do so, insists that the style of argument by analogy offered by Reid is inappropriate to Darwin's argument.[15] Instead, relying on an article by Julian Weitzenfeld,[16] he advocates basically the same form of argument by analogy that we are arguing for in this book. Weitzenfeld first shows two simple games have an isomorphic structure, and then shows how from a winning strategy in one game you may infer a corresponding winning strategy for the other. Unfortunately for Waters' purposes, this argument

<hr/>

[15] Waters (1986), pp. 502–513. Mention here should also be made to Sterrett (2002) as one of the very few commentators on the *Origin* known to understand analogy, and argument by analogy, in its original, classical sense. We do not, however, agree with Sterrett's leading suggestion that in Darwin's analogy, methodical artificial selection is analogous to the principle of divergence in nature, and unconscious artificial selection to the principle of extinction.

[16] Weitzenfeld (1984), pp. 137–149.

falls squarely within the *mathematical* use of analogical reasoning, where the soundness of the argument can be guaranteed purely a priori. However, when Waters writes 'He (Darwin) tried to infer that the result of nature making selections over many generations would be significant modification. He based this inference on the assumption that the structure determining the modification of organs was isomorphic to its natural counterpart',[17] the difference between the use of analogy within mathematics and in its empirical application emerges. In the mathematical case, the winning strategy in the one game can be inferred directly from the isomorphic mapping of the one game onto the other, but in the empirical case, we need additional, further empirical, grounds for supposing that a particular feature of the model can be transferred to its target. This means that there is here a gap in Waters' presentation: we need to specify what Mill called the 'material circumstance', the reason that justifies the transfer of a specific feature from the model to its target.

To discover why this transfer is justified we need to return to see precisely what happens in domestic breeding practices. When human beings select which creatures to promote, they always do so *systematically* – persistently and consistently – always tending to select creatures that possess the *same* features through successive generations. The race horse owner will always tend to send the fastest horses to stud, while farmers will always tend to select the sturdiest. It is precisely this selection of the *same* characteristics through successive generations that is found to reinforce the presence of those characteristics, making them more and more pronounced until they are sufficiently pronounced to constitute a new variety of the species in question. This therefore is Mill's 'material circumstance'.

When we turn to the situation in nature, we will find the same systematicity in the way the struggle for existence discriminates among creatures. In a given environment the same set of characteristics will be favoured through successive generations: the desert will always favour drought-resistant plants, it will always be the best camouflaged mice that tend to escape the predatory birds, and so on. It is precisely this feature that it has in common with the domestic situation that justifies us in inferring that it is possible for NS to be responsible for the formation of new varieties.

Can Natural Selection Be Responsible for the Formation of New Species?

So far, the argument has established that since AS in a domestic context has produced new varieties, so the struggle for existence will have been able

---

[17] Waters (1986), p. 510.

to produce new varieties in the wild. However, this conclusion falls short of what Darwin wishes to show: he wants to show that it is possible for NS to be responsible for the generation of the tree of life and hence the vast changes involved. When we informally illustrate the way NS works, we do so with simple examples such as explaining how camouflage will develop. But such simple examples fall far short of the kind of changes involved in the development from simple organisms to the highly complex creatures we see around us. We need to explain the possibility of the emergence from such simple creatures of the animals we see around us with pancreas, liver and kidneys. In particular, it needs to be shown that it is possible that NS should achieve the kind of irreversible change constituted by the formation of a new species. Darwin needs to show that apparent limitations on the effects achieved by AS need not apply to NS. He needs to address the problem posed by reversion: the apparently universal fact that fresh varieties created by AS were not necessarily permanent changes. If a domesticated animal was released into the wild, its offspring would eventually revert to its feral state, undoing whatever work humans had achieved by their breeding practices.

> Having alluded to the subject of reversion, I may here refer to a statement often made by naturalists – namely, that our domestic varieties, when run wild, gradually but certainly revert in character to their aboriginal stocks. Hence it has been argued that no deductions can be drawn from domestic races to species in a state of nature.[18]

Despite the fact that Darwin clearly attached importance to the difference between unconscious and methodical selection, few commentators have paid any attention to it. It is a merit of Susan Sterrett that she takes it with full seriousness (Sterrett (2002)).

Whatever unclarity may surround the concept of a species, it is clear that the emergence of a new species constituted a permanent change. Therefore Darwin must now argue that the limitations that affect AS may not also apply to NS.

First, here is Darwin's own presentation of the argument in chapter IV of the *Origin*:[19]

> As man can produce and certainly has produced a great result by his methodical and unconscious means of selection, what may not nature effect? Man can act only on external and visible characters: nature cares nothing for appearances, except in so far as they may be useful to any being.

[18] Darwin (1859), p. 14   [19] Ibid., pp. 83–84.

She can act on every internal organ, on every shade of constitutional difference, on the whole machinery of life. Man selects only for his own good; Nature only for that of the being which she tends. Every selected character is fully exercised by her; and the being is placed under well-suited conditions of life. Man keeps the natives of many climates in the same country; he seldom exercises each selected character in some peculiar and fitting manner; he feeds a long and a short beaked pigeon on the same food; he does not exercise a long-backed or long-legged quadruped in any peculiar manner; he exposes sheep with long and short wool to the same climate. He does not allow the most vigorous males to struggle for the females. He does not rigidly destroy all inferior animals, but protects during each varying season, as far as lies in his power, all his productions. He often begins his selection by some half-monstrous form; or at least by some modification prominent enough to catch his eye, or to be plainly useful to him. Under nature, the slightest difference of structure or constitution may well turn the nicely-balanced scale in the struggle for life, and so be preserved. How fleeting are the wishes and efforts of man! how short his time! and consequently how poor will his products be, compared with those accumulated by nature during whole geological periods. Can we wonder, then, that nature's productions should be far 'truer' in character than man's productions; that they should be infinitely better adapted to the most complex conditions of life, and should plainly bear the stamp of far higher workmanship?

It may be said that natural selection is daily and hourly scrutinising, throughout the world, every variation, even the slightest; rejecting that which is bad, preserving and adding up all that is good; silently and insensibly working, whenever and wherever opportunity offers, at the improvement of each organic being in relation to its organic and inorganic conditions of life. We see nothing of these slow changes in progress, until the hand of time has marked the long lapses of ages, and then so imperfect is our view into long past geological ages, that we only see that the forms of life are now different from what they formerly were.

This passage, although it is only two paragraphs long, is crucial to the case Darwin is building. What he does here is explore the analogy between the breeding practices of human beings and the operation of the struggle for existence. What we have is a series of sentences in which Nature is metaphorically personified as a farmer, say, engaged in a programme of breeding in the wild, each sentence comparing the activity of nature with the corresponding activity of human farmers. Every sentence establishes a contrast between human beings and Nature: Nature in a wide variety of respects will prove to be a more efficient and powerful breeder than man. For instance, in addition to the obvious contrast in the vastly greater time that Nature has operated, we have 'nature cares nothing for appearances':

whereas human beings can only select for such favourable variations as are readily apparent,[20] the struggle for existence will select for any favourable variations that are actually there. Even variations that are so slight as to be invisible to humans will confer a slight advantage in the struggle for existence. 'He (man) does not rigidly destroy all inferior animals, but protects during each varying season . . . all his productions.' Human beings engaged in selection do not consistently and ruthlessly discriminate between creatures displaying favourable and unfavourable characteristics, but NS will always and everywhere operate whenever a favourable or unfavourable variation appears. At the same time, Darwin had already written:

> Domestic races of the same species, also, often have a somewhat monstrous character; by which I mean, that, although differing from each other, and from the other species of the same genus, in several trifling respects, they often differ in an extreme degree in some one part, both when compared one with another, and more especially when compared with all the species in nature to which they are nearest allied.[21]

As a result of AS, animals will frequently acquire characteristics that are advantageous to humans, but disadvantageous to the animal. In exchange for these disadvantages, they will be removed from the struggle for existence, and looked after by humans. As a result of this fragile bargain, they will become dependent on humanity for their continued existence. By contrast, 'can we wonder, then, that nature's productions should be far 'truer' in character than man's productions; that they should be infinitely

---

[20] It is worth noting in this connection that it can often be virtually impossible for human beings to determine whether a given trait is, or is not, advantageous, particularly in cases where it would involve a complex cost–benefit analysis to answer the question, leading to some of the disastrous mistakes of artificial selection. Consider a case discussed by Immanuel Kant in *Anthropology from a Pragmatic Point of View*: babies crying.

What could nature's intention be here in letting the child come into the world with loud cries which, *in the crude state of nature*, are extremely dangerous for himself and his mother? For a wolf or even a pig would thereby be lured to eat the child, if the mother is absent or exhausted from childbirth. . . . One must therefore assume that in the first epoch of nature with respect to this class of animals (namely, in the time of crudity), this crying of the child at birth did not yet exist; and then only later a second epoch set in, when both parents had already reached the level of culture necessary for *domestic* life, without our knowing how, or through what contributing causes, nature brought about such a development. (Kant (1798), p. 423 (Ak. 7:327))

However difficult it would be for humans to determine the point at which it would be advantageous for babies to cry, this would be completely straightforward for natural selection. *Solvitur ambulando.* It simply wipes out babies that cry too early, and allows babies that cry as soon as it is safe enough to do so to prosper.

[21] Darwin (1859), p. 15.

better adapted to the most complex conditions of life, and should plainly bear the stamp of far higher workmanship?'

And so on: in these paragraphs Darwin draws attention to the wide variety of different ways in which NS will prove to be an altogether more efficient and powerful agent for change than AS, leading to a picture of Nature as an agent that, operating for a time that is vastly in excess of the time that human beings have existed, selects with utter consistency in a way that leads to a gradual accumulation of small advantages and that simultaneously attends to all the parts of animals and plants and the way those parts interrelate. This is a picture of a selector that is vastly superior to any human selector. And, Darwin argues, if the selector is vastly, even if only quantitatively rather than qualitatively, superior then the changes that such a selector will bring about will be vastly superior to any we can achieve.

What this means is that, given the superiority of NS, whatever limitations there may be to the achievements of AS, these need not apply to NS. It is therefore possible that NS should bring about the vast changes required to generate the tree of life, including effecting the kind of irreversible changes constituted by the emergence of new species. Intuitively this is a strong argument. What we need is a theoretical understanding of the way that the possibility of making a stronger claim for the power of NS than could be made for the power of AS is a natural development of the kind of argument by analogy that we have ascribed to Darwin so far.

A Fortiori

There is actually a widespread precedent for the way Darwin argues here within the theology of Abrahamic faiths: we start by drawing an analogy between God's person and activities and human beings and their activities.[22] We then argue that given the superiority of God over anything within the created world, what God can achieve vastly outstrips anything that human beings can achieve. What we find in the *Origin* at this point is a meeting of Athens and Jerusalem, transposed into a naturalistic, modern, context, with the argument that concerns us replacing God by the natural environment interacting with creatures.

---

[22] We already looked in Chapter 2 (section "William Paley, *Natural Theology*") at a strand in Paley's *Natural Theology* that fits this pattern.

One of the very few authors to pay attention to the argument of chapter IV of the *Origin* by comparing that argument to the way such arguments have recurred in theological discussions was H. Bartov.[23] We may clarify Darwin's argument further by looking at the way in which Bartov makes the wrong comparison at this point. He argues that the central arguments of the *Origin* both for a struggle for existence in the wild and for the capacity of NS to produce new species are a fortiori arguments and therefore not arguments by analogy. Unfortunately, he has been misled by an ambiguity in the characteristic formula 'how much the more' and as a result an ambiguity in the notion of an 'a fortiori argument' that is relevant to Darwin's work. As a consequence Bartov misses the central point of the argument in chapter IV we are looking at.

Bartov does not say why he believes there is an incompatibility between an argument by analogy and an a fortiori argument. In fact, on the usual understanding of the phrase, an 'a fortiori argument' is not the name of a specific form of argument; it is simply an argument in which the reason we have for accepting the premise gives us an even stronger reason for accepting the conclusion. Arguments of any logical form whatever can, on this understanding, with appropriate subject matter be a fortiori.

Bartov explains his understanding of an a fortiori argument as follows: 'The *a fortiori* consists in showing that a certain conclusion, proved to be true in one instance, is also true in another one, in which there are stronger and/or more reasons for its truth, than in the first one.'[24]

Accordingly, he renders Darwin's argument as follows:

— Man produces new varieties and species by Artificial Selection.
— Natural Selection is more efficient in changing varieties and species than Artificial Selection.
— Therefore, Natural Selection *certainly* produces new varieties and species.[25]

But that is not Darwin's argument: the first premise is not Darwin's, and the conclusion is not the one he is after. As far as Darwin knew, Man did *not* produce new species by AS, and the point of Chapter IV is not simply to show that it is more certain that NS could do the same as AS, but to show that NS could do significantly more than AS.

If we consider a typical religious formula, 'If humans can F, how much more can God F', this can be interpreted in two different ways: either as saying that if humans can F, it is even clearer that God can do the same, or

---

[23] Bartov (1977), pp. 131–145.   [24] Ibid., p. 131.   [25] Ibid., p. 139.

as saying that if humans can *F*, God can *F* with greater power. Bartov has opted for the first reading, but it is the second that provides the appropriate parallel for Darwin's argument. If we do call Darwin's argument an 'a fortiori' argument it is according to the second reading of the 'how much the more' formula.

A true parallel for Darwin's argument can be found in the letter to the *Hebrews* in the New Testament, where, despite the completely different, and to our eyes somewhat strange, subject matter, the pattern of argument closely follows that of Darwin's. *Hebrews* is a book that is structured so as to show figures from the Old Testament and their religious practices as foreshadowing the life and work of Jesus Christ but at the same time falling far short of Christ, and thus failing to achieve what he achieved through his ministry. *Hebrews* puts at the centre of its argument the case of the Old Testament priests giving sacrifices for the forgiveness of sins. The priests are represented as themselves weak and sinners, whose sacrifices are as a result incapable of achieving a complete reconciliation of God and His people, needing to be repeated daily. By contrast, Christ is presented as capable of achieving what they could not:

> *Heb.* 8, 6 Christ has obtained a ministry which is as much more excellent than the old as the covenant he mediates is better.

> *Heb.* 26 For it was fitting that we should have such a high priest, holy, blameless, unstained, separated from sinners, exalted above the heavens. 27 He has no need, like those high priests, to offer sacrifices daily, first for his own sins and then for those of the people; he did this once for all when he offered up himself.

Here, we first set up the Old Testament priesthood as an analogy for the ministry of Christ ('a copy and shadow', *Heb.* 8, 5). We then stress the inferiority of the priests and their activities to Christ and His activities. We then conclude that He can achieve what they could not. Despite the utterly different subject matter and context, we clearly have here exactly the same structure of argument as that which we found in *Origin*, chapter IV.

Alternating the Analogy

The argument of chapter IV of the *Origin* consists in a further exploration of the analogy between AS and NS. Paradoxically, however, it is concerned here not with further points of similarity between what happens on the farm and in the wild, but with a series of contrasts. So the question arises,

'Are we still talking about an argument by *analogy*, if a crucial stage of the argument consists in contrasting the two cases?' R. A. Richards,[26] who we discuss in detail in Chapter 7, argues that the fact that Darwin is seeking to make a stronger claim about the power of NS than the power of AS ipso facto shows that we are no longer talking about analogy. Richards understands 'argument by analogy' as the form of argument we have ascribed to Thomas Reid: if two situations, A and B, share a number of properties, every property that A shares with B increases the likelihood that they also share some further property of A that concerns us. On this understanding, it is clear that in chapter IV of the *Origin* we are no longer talking of such an argument. However, here there emerges a further major difference between Reid's account and the classical argument by analogy that we have ascribed to Darwin's use of analogy. What we have here is a natural development of the classical argument.

Against this background, we can now set out the structure of the whole argument. We start with a vast array of simple analogies along the lines 'This farmer is to that bull as this fox is to that rabbit', building up to an analogical model, 'the entire community of human breeders is to animals and vegetables under domestication as Nature (the natural environment interacting with creatures in the wild) is to creatures in the wild'. From there we argue by analogy to the first conclusion: The community of human breeders is to the generation of new varieties under domestication as Nature is to the generation of new varieties in the wild.

The next step is to argue that human beings are inferior selectors to Nature. The limits on human powers are stressed, and Nature is seen in a wide variety of ways to be a vastly more efficient selector acting over a vastly longer time. This means we can see that although AS can provide an analogical model for NS it is a vastly inferior model, and we can see that what happens in Nature is a vastly more powerful version of what happens under domestication, 'and consequently how poor will [Man's] products be, compared with those accumulated by nature during whole geological periods'.

What has been done here is that a fundamental feature of analogy has been exploited: that analogies alternate ('A is to B as C is to D implies that A is to C as B is to D'). If the human community is to the generation of new varieties under domestication as Nature is to the generation of new varieties in the wild, then the human community is to Nature as the generation of new varieties under domestication is to the generation of new

[26] Richards (1997).

varieties in the wild. What alternating the analogy does is let us see the world of Nature as the farm scaled up and everything that happens in Nature as that which happens on the farm writ large.

This gives us the conclusion that Darwin wants: the more powerful the selector, the more powerful the changes wrought by selection. Hence the limitations on the success of AS need not apply to NS, and it is possible that NS should be responsible for the vast changes required to create new species and even the whole tree of life.

CHAPTER 6

Darwin's Use of Metaphor in the Origin

With the analysis of Darwin's argument by analogy from artificial to natural selection in the first four chapters of the *Origin*, we have completed our main task. In this chapter and the next, we look at two consequent issues in the light of Darwin's use of this analogy. In Chapter 7, we shall reply to some authors who have argued for different interpretations of the argument with which we are concerned, and in this chapter we shall examine something for which Darwin was frequently criticised, or misunderstood, at the time – the extensive use of metaphor throughout the *Origin*, especially metaphors that represent Nature as a farmer engaged in a vast breeding programme concerning the whole animal kingdom.

There is an extensive literature on the uses of metaphor in science. It is possible to identify two main strands in that literature. First, there is the widely influential complex of ideas developed by Mary Hesse.[1] Here, and on the part of those who have followed in her footsteps, it is argued that a number of scientific theories are not so much straightforward literal descriptions of the phenomena as in fact presenting us with analogical models of those phenomena. As a result a large number of the claims made by those theories are to be regarded as metaphors based on the analogies underlying the models. Second, quite independently of Hesse's concerns there is a much simpler idea which is uncontroversial and almost universally accepted. At times of new scientific discovery, the discovery will frequently take the form of being struck by an analogy – the heart as a pump, etc. As a result at that stage of the scientific process, it will only be possible to offer metaphors instead of a straightforward literal theory. In the second strand, unlike the first, the metaphors are seen as used in an

[1] See Hesse (1963). For an analysis of analogy and metaphor in the *Origin* that draws on Hesse's work, see Beer (1983), ch. 3.

essentially *provisional* stage of the scientific enquiry. These metaphors are to be replaced eventually by a precise statement that has dispensed with metaphor, possibly by giving to the words that were originally used metaphorically a new technical meaning.

We do not dispute the significance of Hesse's work for the philosophy of science. The use of artificial selection as an analogical model for natural selection, however, is quite different from that envisaged by her work. Whereas she is concerned with cases where a scientific theory takes the form of presenting an analogical model of the phenomena, in the case of the *Origin* we are concerned with Darwin's defence of a theory which he had already arrived at, as had Wallace,[2] prior to his being struck by the analogy. Here we have a case of a theory that it is perfectly possible to present without any reliance on the analogy. The use of artificial selection as an analogical model is not then to give Darwin a means of presenting his theory, but it is rather to provide empirical support for that theory by showing in an empirically accessible environment that the repeated discrimination in favour of favourable traits can lead to the generation of fresh varieties. As we shall see, he will use metaphor as a way to develop and explore the analogy ever further. This is a use of metaphor that is completely different from that proposed by Hesse, and indeed in the way that it is tailored to Darwin's project, possibly without parallel in the rest of science.

There are certainly some examples of Darwin using metaphors to form new concepts that are appropriate for his purposes – this is particularly true with his metaphorical use of 'struggle'. It is intended that *such* uses of metaphor will eventually be replaced by literal statements, possibly by giving the word used metaphorically a settled technical sense.[3] However, what primarily concerns us are the metaphors based on the analogy that is the principal concern of this book: metaphors that describe Nature as if she were a farmer engaged in a vast breeding programme covering the whole animal and vegetable kingdoms. These are the metaphors that give to Darwin a simple and perspicuous way to explore the analogy between artificial and natural selection. They are also far and away the richest and most complex of Darwin's metaphors.

---

[2] Wallace (1858).

[3] Cf. Darwin (1859), p. 485: 'The terms used by naturalists of affinity, relationship, community of type, paternity, morphology, adaptive characters, rudimentary and aborted organs, &c., will cease to be metaphorical, and will have a plain signification.'

Metaphor and Analogy

Although the concepts of metaphor and of analogy are frequently confused, they should be clearly differentiated. An analogy is not a linguistic matter at all, but a relationship existing in the world: that there is an analogy between natural selection and artificial selection is a matter of fact; it is a four-termed relationship existing in nature, in the same way that the fact that 4 is to 2 as 34 is to 17 is a fact of mathematics, not language. That such analogies exist can be expressed by a wide variety of linguistic means, including metaphor, but the analogy is not a piece of language but what the language is about. On the other hand, we shall restrict the term 'metaphor' to specific pieces of language, quotable passages, such as 'nature cares nothing for appearances'.

Darwin's use of metaphor proved controversial, with many people being misled into thinking that his talking about 'natural selection' implied that he was actually thinking of Nature as an agent with purposes, and some, who were not so misled, thinking that others might be. In particular Wallace wrote to Darwin[4] to persuade him to adopt Herbert Spencer's phrase 'survival of the fittest' instead. Although Darwin initially acquiesced in the suggestion,[5] this was arguably a mistake. Whereas the metaphor 'natural selection' perfectly encapsulates the whole theory in a single phrase, 'survival of the fittest' only captures one component of the theory, failing to include crucial ideas such as that of hereditary variation.

Being misled by Darwin's metaphors into thinking that he was regarding Nature as having intentions is a consequence of failing to grasp the point that Kant makes concerning 'cognition according to analogy':

> [W]e do not attribute to the Supreme Being any of the properties by which we represent objects of experience in themselves, and thereby avoid dogmatic anthropomorphism; but we ascribe them to His relation to the world, and allow ourselves a symbolical anthropomorphism, which in fact concerns language only, and not the object itself.[6]

---

[4] Letter to Darwin, 2 July 1866.
[5] Cf. Darwin (1868), p. 6: 'This preservation, during the battle for life, of varieties which possess any advantage in structure, constitution, or instinct, I have called Natural Selection; and Mr. Herbert Spencer has well expressed the same idea by the Survival of the Fittest. The term "natural selection" is in some respects a bad one, as it seems to imply conscious choice; but this will be disregarded after a little familiarity.'
[6] Kant (1783) §57.

For Kant, the *language* we use to talk about God is unavoidably anthropomorphic, but once we grasp that we are only talking about God *by analogy* we avoid all taint of anthropomorphism by restricting our interpretation of that language to only what can be justified by the underlying analogy. Thus, when we say that the farmer is to the bull as the frost is to the daisy and describe them both as selecting, we are only talking about their respective relations to the bull and the daisy, without thereby implying that the farmer and the frost share any further intrinsic properties.

So too, we will not be misled by Darwin's metaphors if we restrict their interpretation to what can be justified by the analogies on which they are based. We will then see that the struggle for existence is not just a matter of Tennyson's 'nature red in tooth and claw' and that there is nothing anthropomorphic entailed by the metaphors of selection. Darwin was perhaps unduly optimistic in assuming a readership that would realise this.

Our task in this chapter is to look more closely at the relation between the concepts of analogy and of metaphor, and to explain in what ways recourse to metaphor serves Darwin's purposes in the *Origin*.

Before turning to look in detail at the uses Darwin makes of metaphor, there are some general observations about metaphor that bear directly on those uses. We need first a clear understanding of the relation of metaphor to analogy. In the *Poetics*, Aristotle divided metaphors into four types, of which the first three were variations of the idea that what justified a metaphorical comparison between two things was simple similarity, the possession of common properties; but the fourth type was metaphors based on analogy. In the *Rhetoric*, he then maintains that the most successful metaphors were those based on analogy and not on simple similarities.[7] However, although he gives an excellent range of examples to illustrate this contention – both positively and negatively[8] – he gives no theoretical explanation of the fact.

Aristotle's lead here has been largely neglected in the history of discussions of the theory of metaphor. One author who does take it up is Paul

---

[7] See, *Rhet.* 1411a 1ff.

[8] E.g., the Pythagorean metaphor, not based on analogy: 'It is, *e.g.*, metaphor to call a good man "square", since both are perfect, but it is lifeless' (*Rhet.* III, 1412a 1). This is contrasted with a metaphor based on analogy: 'And as Homer often uses it, animating the inanimate by a metaphor. His fame derives from the way he makes things actual in all his work. Thus, *e.g.*, .., "Many of the spears stuck in the ground, straining to gorge themselves on flesh."' (*Rhet.* III, 1412a 3).

Henle, who, in an excellent article,[9] not only gives analogy a central role in his whole discussion but offers a clear explanation wherein the superiority of metaphors based on analogy lies.[10] One of Henle's main theses is that metaphor is unique among the figures of speech in being extensible: given an initial metaphor we can follow it with further metaphors that develop the comparison yet further, and that this provides the answer to the question wherein the superiority of metaphor lies: 'the aptness of metaphor depends on the capability of elaborating it – of extending the parallel structure.'[11] This point is then immediately linked to analogy: 'Metaphor . . . can be spun out, following a line of analogy, or even several lines at once, carrying it quite far.'[12]

To see how this works, we need to look in detail at metaphors based on analogy.

Metaphors Based on Analogy

In the *Poetics*,[13] Aristotle gives us an early attempt at a classification of figures of speech, all of which at this stage he calls metaphors. These are all figures in which a word is transferred from its proper application to something to which it cannot be literally applied. It is a strange classification, apparently more driven a priori than by any survey of the phenomena. This account is both broader and narrower than what we would now understand by the term 'metaphor'. It is broader in that it will include many examples which we would see as quite different figures from metaphor, and it is narrower in that it only concerns cases of metaphor where it is a single word in a sentence that is being used metaphorically.

He here devotes far more space to metaphors based on analogy (μεταφοραί κατά τό ἀνάλογον) than any other metaphor. He does not, however, explain why. The metaphors based on analogy in the *Poetics* all

---

[9] Henle (1958).
[10] We may also mention a clear and insightful treatment of metaphor and analogy in Ferguson (1792). His main theme, as in Aristotle, is the capacity of metaphors based on analogy to draw comparisons between things that are unlike ('a bird may be said to swim through the air'). (In this connection we may note for our purposes that Darwin's use of metaphor involves such comparisons as that between a racehorse owner and a severe frost, on the basis that the racehorse owner favours the best runners, while the frost favours the hardiest plants.) Also of relevance to our concern is Ferguson's remark, 'Allegory may thus be considered as metaphor extended from single qualities to many such.'
[11] Henle (1958), p. 180.   [12] Ibid., p. 181.   [13] *Poe* 1457b, 6–33.

follow a simple linguistic schema: 'I call 'by analogy' cases where A is to B as C is to D: one will then say C instead of A, or vice versa. People sometimes add as follows: 'A is the C of B'.[14] So either, say, calling old age 'evening', or 'the evening of life'. It is at this stage completely straightforward to see in what way such metaphors are based on analogy and not on simple similarity. He will finally claim[15] that metaphor is the most important linguistic device in poetry, but does not elaborate. All in all the metaphors he actually uses to illustrate his account are typically trite.

When we look at the *Rhetoric*, although he begins by claiming to adopt the account of metaphor he has outlined in the *Poetics*, the contrasts between the two works are striking. All the metaphors he cites as examples are cases of what we would unhesitatingly call metaphors: it is in the *Rhetoric* that our concept of metaphor is established. The metaphors used are typically not the simple single-word metaphors of the type outlined in the *Poetics* but of far greater complexity. And the metaphors are also typically of far greater interest: the whole discussion comes alive.

What concerns us here is the claim which pervades his discussion that the successful metaphors are not based on simple similarity but on analogy,[16] where, given the context in which he is writing he primarily means rhetorically successful. For our purposes the following questions arise. Given that the metaphors claimed to be based on analogy frequently have a far more complex linguistic structure than those in the *Poetics* and do not conform to the simple schema outlined in the *Poetics*,[17] in what way can they nevertheless be thought of as resting on proportional analogies? What is the form of such metaphors? And why is it that it is such metaphors that are successful?

Let us consider an example:

> Many of the spears stuck in the ground, straining to gorge themselves on flesh[18]

---

[14] *Poe* 1457b, 19–20.   [15] *Poe* 1459a, 5

[16] *Rhet.* 1411a 1. The only time he cites a metaphor that is not based on analogy – calling a good man square – it is to illustrate his contention that such metaphors are wooden. (1411b 20).

[17] In the only case in which Aristotle actually spells out the underlying analogy, he does so in a way that is quite different from the way envisaged by the *Poetics* schema: '"Again the ruthless stone rolled down to the plain"... Homer has attached these attributes by analogy; for the stone is to Sisyphus, as the impudent person is to the one he cheeks.' (*Rhet.* 1412a, 2–3).

[18] *Rhet.* 1412a 1, citing *Iliad* xi 574.

This is taken by Aristotle from the scene in the *Iliad* where Ajax is singlehandedly warding off an attack by the Trojans who are bombarding him with spears, where the spears are failing to hit their target and either lodge in Ajax' shield or fall short. The metaphor is cited by Aristotle as an example of Homer presenting the inanimate as animate and 'making us see things'. In this, despite its being highly fanciful it succeeds admirably.

We characteristically interpret a metaphor such as this immediately, but for our purposes we need to spell out the process of interpretation. We begin by asking 'What would, in this context, unlike spears be actually "straining to gorge themselves on flesh?"', with a natural answer being hunting dogs that have been leashed in, baulked of their prey. This leads to the beginning of an answer: the metaphor is initially based on the analogy *spears are to Ajax as hunting dogs are to their prey*. But to make sense of the detail of the text, we have to see that initial analogy having been amplified by a series of subsidiary analogies: *being stuck in the ground is to spears as being leashed in is to dogs*; *wounding Ajax is to spears as gorging themselves on flesh is to dogs*; *quivering in the ground is to spears as straining on the leash is to dogs*, and possibly yet more. In this way, we have arrived at the idea of a metaphor not based on a single analogy, but on a network of interrelated analogies. These analogies combine to allow us to see the situation of the dogs on a leash as an analogical model of the spears falling short of their mark.

Because the *Rhetoric* is primarily a practical, rather than a theoretical, work, giving instruction to orators, there are at most hints as to the answers to many of the questions that this account gives rise to. In particular, what is a full account of the way such metaphors are formed, and why is it these metaphors that are the successful ones? On the second question, Aristotle cites characteristics that make these successful: "they make us see things", "they bring things to life (have ἐνέργεια)', are witty (ἀστεῖα), but it requires further explanation as to why this is so.

We can derive from Aristotle's treatment of metaphor in the *Rhetoric* the following 'programme' for the study of metaphor. (1) Always understand by 'analogy' proportional analogy – A is to B as C is to D. (2) Concentrate on metaphors based on analogy, stressing that it is such metaphors that are the significant and successful ones – however we gloss the idea of success. (In Aristotle's own case, what is successful is determined empirically: these metaphors are the ones that have proved popular.) (3) Engage in the extensive concrete and detailed exploration of highly complex metaphors.

But although Aristotle's *Rhetoric* is an excellent starting point for our purposes, we need to go beyond him if we are to give a theoretical account of where this programme leads. Unfortunately his central stress on the importance of metaphors based on analogy has been widely ignored with relatively few authors even considering the claim.[19]

Once we engage with metaphor in the way outlined, we arrive at a particular account of the way such metaphors are formed. Consider a simple case – Ferguson's '*A bird swims through the air*'. Here we have a metaphor based on the analogy *a bird is to the air as a fish is to the sea*, and the metaphor compares two situations – the actual situation of a bird flying through the air with the fish swimming through the sea. The metaphor is formed as follows: we juxtapose two sentences describing these two situations:

A bird flies through the air.
A fish swims through the sea.

We form the metaphor by constructing a new sentence that selects each of its words from one of these two sentences, in such a way that the alert reader can reconstruct the two situations being talked about, the actual situation and the one it is compared to.

A bird *swims* **through the air.**

With a simple example like this it is hard to show why this kind of metaphor can be such a powerful linguistic tool – for that one has to look at complicated comparisons where it is between a situation and an analogical model that involves not a single analogy but multiple analogies (in the case that interests us, the model involves comparing several instances of farmers choosing bulls, with the weather affecting several plants).[20]

---

[19] The extent of the blind spot is illustrated by Kirby (1997). In the section of this article devoted to metaphor in the *Rhetoric*, Kirby only mentions analogy once, and that is an entirely incidental reference. The central Aristotelian claim that it is precisely metaphors based on analogy that are the successful ones is not even mentioned, ascribing claims that Aristotle makes about such metaphors to metaphors in general, including the ones Aristotle contrasts with analogical metaphors.

In recent times, we may cite three authors who *do* follow what we have just called Aristotle's programme: Paul Henle, who we have already just looked at; Nowottny (1962) and White (1996). All three of these arrive independently at versions of the account of metaphor we shall present here. We may mention in connection with Nowottny's work Leech (1969), where the section on metaphor is presented as a formalised presentation of her account. We should also mention Dedre Gentner, who has explored the relation of metaphor and analogy, but largely as a psychological study of the role of analogy in thought, making her work less directly relevant in our context.

[20] Aristotle himself seems to be indicating a similar conception at *Rhet.* 1412b 11: 'Similes too ... are always as it were successful metaphors, since they are always said in two different ways (ἐκ δυοῖν

A major reason why it is that metaphors based on analogy are the successful and interesting ones is that stressed above all by Henle: these are the metaphors that are extensible. It is also the feature of such metaphors that is directly relevant to Darwin's use of metaphor. Metaphors by analogy can be a starting point, opening up to us a wide range of related ideas. Consider here the following example Aristotle cites from Isocrates' *To Philip* §127:

> While it is only natural for the other descendants of Heracles, and for men who are under the bond of their polities and laws, to cleave fondly to that state in which they happen to dwell, it is your privilege, as one free to roam at large (ἀφετός), to consider all Hellas as your fatherland.

At a time when Greece was confronted by a threatened Persian invasion, Isocrates in *To Philip* is urging Philip of Macedon to unite and lead the Greeks to meet the threat. He portrays Philip as the one leader who was capable of adopting the larger view while the other Greeks were squabbling and exclusively preoccupied with the concerns of their own city states. It is here that he introduces the metaphor 'free to roam at large (ἀφετός)'. The word 'ἀφετός' was the term used to designate an animal that had been dedicated for sacrifice: once an animal had been so dedicated it was sacred and it was forbidden to interfere with or molest such an animal. While other animals were circumscribed and forbidden to trespass outside their own fields, the sacred animals could wander at will. Here Isocrates evokes and encapsulates an entire political situation by the metaphorical use of a single word, with Philip's cosmopolitan outlook being compared to being dedicated to the gods, the other leaders' petty preoccupation with their own city states to animals unable to move freely, and Philip's freedom of movement to that of the sacred animal. What we need to understand in detail is the way that by parity of reasoning, seeing nature as a superhuman farmer can evoke and encapsulate the whole theory of evolution by natural selection.

Informally, the way that metaphors can be extended is easy to understand. Suppose that we have a metaphor comparing A with C, on the basis of an analogy between A and C (for some B and D, A is to B as C is to D).

λέγονται), like metaphors based on analogy', citing as an example Thrasymachus comparing Niceratus to 'Philoctetes stung by Pratys' 'when he saw Niceratus, defeated by Pratys in the rhapsodes' competition, still dirty with his hair uncut'. (Philoctetes stumbling into a sacred grove had been stung by a snake, so that not only is Niceratus in his dishevelled state compared to the wounded Philoctetes, but Pratys is compared to a poisonous snake.)

We can then produce a succession of further metaphors in which we describe A further, but talking as if A were C. Thus in *Antony and Cleopatra*, Shakespeare, having already used metaphors based on the analogy *Antony is to Cleopatra as the Sun is to the Moon*, when Antony dies makes Cleopatra describe this in terms of a sunset: 'there is nothing left remarkable/Beneath the visiting moon'.

We shall see in detail how this works when we look at the way in which Darwin exploits this feature of analogical metaphor, and when we sketch a brief theoretical account of the way in which it is possible to extend a metaphor with yet more metaphors that can be regarded as developments of that initial metaphor. The key point is that it is possible to develop a simple analogy 'A is to B as C is to D' into an analogical model, involving multiple analogies. This is done by correlating features of A and its situation with corresponding features of C and its situation, and setting up analogies between these features. Suppose that we compare A and C. In working out this comparison, the way that A* is related to A in A's situation corresponds to the way that C* is related to C in C's situation and hence we say A* is to A as C* is to C, and form a secondary comparison between A* and C*. (If Antony is to Cleopatra as the Sun is to the Moon, then Death is to Antony as Sunset is to the Sun). So, from an initial comparison we may form one or more secondary comparisons, and this in turn is a clearly reiterable process. Once we have established such an analogical model, we may then form metaphors that describe A and its situation as if we were talking about C and its situation. (Once we have used artificial selection to model natural selection, we can talk about Nature's discrimination between creatures in the wild as if we were talking about a domestic breeder discriminating between domestic animals and plants.)

Scientific and Poetic Uses of Metaphor

The literary critic George Levine writes:

> The first shock of my first serious reading of *On the Origin of Species* was the discovery that it was written in the voice of a man who allowed himself to express the feelings nature aroused in him, and who sought the richest metaphorical possibilities in his attempt to describe afresh what the world is like.[21]

---

[21] Levine (2011), p. xiv.

and commenting on the opening paragraphs of chapter IV of the *Origin*:[22]

> Having made natural selection a loving and highly moral person, Darwin bursts into almost biblical lyricism when he exclaims, 'How fleeting are the wishes and efforts of man! How short his time!' The whole sequence has the texture not of scientific tract but of literature charged with meanings beyond the literal, and registering nuanced feeling even as it attempts strenuously to make a careful argument.

Levine, and others,[23] have invited us to look at the *Origin* as 'literature', and to celebrate a use of language – including above all Darwin's abundant use of metaphor – comparable with that of poetry or the novel. This praise is, however, misplaced, distracting attention from what is truly impressive about Darwin's use of metaphor. Metaphor has a legitimate use in science just as much as in poetry or rhetoric. However, there are deep differences between the use of metaphor in science and in poetry that are overlooked in these accounts. The poet and the scientist are subject to different disciplines, leading to a number of differences in the use that they may make of metaphor. Before turning to the specific case of the metaphors in the *Origin*, it is worth spelling some of these out.

The poet can show extraordinary flexibility in the interpretation of the analogy underlying the metaphors. As we saw, in *Antony and Cleopatra*, the play is built round metaphors based on the analogy *Antony is to Cleopatra as the Sun is to the Moon*. However, in the course of the play this analogy is given a range of different interpretations, varying from context to context. In some metaphors, Antony is constant, Cleopatra vacillating, in others, Cleopatra is dependent on Antony, and in yet others, Cleopatra derives life and light from Antony. In contrast with this, when Darwin uses a sequence of metaphors based on the analogy between artificial and natural selection, he establishes at the very outset how that analogy is to be understood, and each metaphor is to be interpreted as governed by that basic analogy and is always to be interpreted solely as a legitimate development of that analogy.

Metaphors in poetry tend to resist literal paraphrase: they have an open-endedness that resists reducing to a single prose equivalent. By contrast Darwin's metaphors always make a specific scientific claim that could in

---

[22] Ibid., p. 86.   [23] The best known exponent remains Beer (1983).

principle be spelled out, even if with difficulty, in plain literal language. Even if on occasion it takes effort to recover a prose equivalent and when discovered this prose equivalent is far more cumbersome than the metaphor, it is invariably possible to find it, and there is then absolutely no loss of cognitive content.

Most significantly, the metaphors of the *Origin* are always *functional*, having a specific use in the development of the scientific argument. When Darwin writes, 'How fleeting are the wishes and efforts of man! How short his time!' it is not a matter of 'biblical lyricism', but because it is a vital link in his argument at that point to stress the inferiority of human attempts at selection to those of Nature. It is invariably possible to identify the specific work that each metaphor does in the development of the argument of the *Origin*. We may distinguish three different functions fulfilled by the metaphors of the *Origin*, which we shall look at in turn.

Metaphors as Shorthand

We begin with Darwin's own justification for his use of metaphor. He adds a paragraph in the third edition of the *Origin* defending his use of metaphor, replying to some incautious readers who had read Darwin as seeing natural selection as 'an active power or deity'. He compares his way of speaking with some other scientific uses of metaphor, and then says, 'Every one knows what is meant and is implied by such metaphorical expressions; and they are almost necessary for brevity'.[24] He somewhat tartly concludes the paragraph by saying, 'With a little familiarity such superficial objections will be forgotten.'[25] The metaphors provide a convenient shorthand: even though it is possible to translate his metaphors into prose, the result is characteristically longwinded. Darwin does not explain why this is so, presumably having discovered it by experience, but the reason seems to be an aspect of his central analogy that we have not hitherto commented on: a farmer deciding which animals to allow to breed is a simple and familiar matter that can be described succinctly, but

---

[24] Darwin (1861), p. 85.

[25] Maybe. But note that Jerry Fodor writes, 'Daniel C. Dennett suggests that, if Jones's behaviour is an adaptation, then it's (not Jones but) 'Mother Nature' who is concerned about his contribution to the gene pool. But you might as well blame the Easter Bunny. There isn't any Mother Nature, and if unattached motives can't explain behaviour, neither can the concerns of fictitious persons.' (*TLS*, July 29, 2005, p. 4)

the processes in nature that favour one animal rather than another are complex and varied.

Describing nature as *choosing* which animals shall live and which shall die provides us with a simple epitome of a complicated set of processes. When a pigeon fancier wishes to breed long-beaked pigeons, this is a process that can be described simply by describing a succession of choices in which it is always the longest beaked pigeons that are permitted to reproduce. By contrast, when we consider what is involved in an animal's succeeding in the struggle for existence in the wild, it has to contend simultaneously with a vast range of heterogeneous difficulties – it must struggle against the cold, scarce food resources, predators and diseases, compete with rival members of its own species, and so on.[26] Therefore what counts in the wild as a favourable variation (or, rather set of variations) involves an extraordinarily complex calculation, particularly when we consider that a variation making an animal better able to cope with one difficulty may make it worse able to cope with a different difficulty (for example, the display necessary to attract a mate also makes it more vulnerable to predators).

Because of this, whereas it is simple to describe the whole process whereby a pigeon fancier produces the desired variety of pigeon, it would be humanly impossible to give a full detailed description of the full complexity of what happens in the struggle for existence in the wild. The metaphorical description of the latter process as also one of selection enables Darwin to cut through that complexity to the crucial point that we have here a process that will work in basically the same way as the simple processes of human beings' choosing which animals to breed from.

Metaphors as Extensible

We outlined above the way in which an initial metaphor can be followed by a second metaphor that may be regarded as a development of the idea that justifies the first metaphor. This second metaphor is subordinate to the first. The first metaphor makes perfect sense and may be understood in

---

[26] Cf. Darwin (1859), p. 61 'Owing to this struggle for life, any variation, however slight and from whatever cause proceeding, if it be in any degree profitable to an individual of any species, *in its infinitely complex relations to other organic beings and to external nature*, will tend to the preservation of that individual, and will generally be inherited by its offspring' (emphasis added).

its own right, whereas the second metaphor can only be understood and interpreted on the assumption that we are already seeing Nature as selecting. The second metaphor has to be seen as essentially an extension of the first.

Such metaphors that are extensions of the initial metaphor of Nature as selecting recur throughout the *Origin*, but come into their own in the opening paragraphs of chapter IV. It is here that we see the full utility of Darwin's expressing himself by metaphor. Metaphor provides him with the linguistic means for exploring and elaborating his initial analogy between natural selection and artificial selection. Chapter IV begins by arguing that Nature will prove to be an altogether superior selector to man. By personifying Nature as a woman or goddess ('she'), Darwin is enabled to use a sequence of metaphors to make a series of direct comparisons between natural processes and human activities.

> Man can act only on external and visible characters: nature cares nothing for appearances, except in so far as they may be useful to any being.[27]

A shepherd will choose an inferior sheep provided that it *appears* to have favourable characteristics, but in the wild an inferior sheep is at a disadvantage, no matter how favourable its characteristics may appear to be.

> She can act on every internal organ, on every shade of constitutional difference, on the whole machinery of life.[28]

Because human beings can only select on the basis of a few visible characters, artificial breeding will tend to produce creatures that are distortions from a fully healthy state ('He often begins his selection by some half-monstrous form'), but nature, by operating simultaneously on 'the whole machinery of life' will tend to move from one healthy state to another.

> Man selects only for his own good; Nature only for that of the being which she tends. Every selected character is fully exercised by her; and the being is placed under well-suited conditions of life[29].

Here instructively, Darwin not only compares human selection with natural selection, but registers the limit of the analogy – the point at which the analogy goes lame. Human beings can choose a wide range of different characteristics to select for, according to their desires. Nature is utterly

---

[27] Darwin (1859), p. 83.   [28] Ibid., p.83   [29] Ibid., p. 83.

constrained, and has no choice in the matter, which means that we are not really talking about selection at all, but only about a process analogous to selection and that can be described metaphorically as selection.

> How fleeting are the wishes and efforts of man! How short his time! and consequently how poor will his products be, compared with those accumulated by nature during whole geological periods. Can we wonder, then, that nature's productions should be far 'truer' in character than man's productions; that they should be infinitely better adapted to the most complex conditions of life, and should plainly bear the stamp of far higher workmanship?[30]

This passage could all be regarded as literal were it not for the final ascription to nature's productions as bearing the stamp of 'far higher workmanship', with its implication of skill on the part of nature.

> It may be said that natural selection is daily and hourly scrutinising, throughout the world, every variation, even the slightest; rejecting that which is bad, preserving and adding up all that is good; silently and insensibly working, whenever and wherever opportunity offers, at the improvement of each organic being in relation to its organic and inorganic conditions of life.[31]

This is perhaps the boldest of all these metaphors. The farmer at the market scrutinises the bulls for sale to select the one or two best suited to his purposes, and thus selection is made as a response to a prior scrutiny. Natural processes automatically and instantly confer advantages to some animals in the struggle for life, and therefore can be seen as 'daily and hourly' having 'scrutinised every variation'.

These metaphors are all in the service of comparing artificial selection with natural selection, in order to establish the superiority of natural selection, by showing a wide variety of ways, in addition to the obvious fact that natural selection has a vastly greater time in which to operate, in which natural selection will prove to be a far more efficient selector than Man. By presenting the comparisons in the form of metaphors, and describing Nature as if 'she' were a woman or goddess, Darwin is enabled to make explicit the respects in which Nature outperforms Man: it is as though one is comparing the performances of two human beings – one an amateur and one an expert.

---

[30] Ibid., p. 84.
[31] Ibid., p. 84. For a critique of adaptationist Darwinism which pivots on the rejection of this metaphor, see Milo (2019), discussed in Radick (2019).

All of these metaphors are in perfect accord with the conditions we sketched above for the use of metaphor in science: they are all simply elaborations of the initial analogy between artificial and natural selection; in every case, the content of these metaphors can be derived from reflection on that initial analogy; and in each case it is relatively easy to give a literal gloss on the metaphor, their function merely being to facilitate a direct comparison with the corresponding literal statement concerning artificial selection.

Metaphors and Concept Formation

There is a further vital role that Darwin assigns to metaphor. This is the way in which he exploits metaphor in *concept formation* and the creation of new, literal, meanings. Quintilian pointed out that one use of metaphor was 'out of necessity'. There are cases where we need to talk about something but lack the means to give a literal description of the phenomenon in question. We may then resort to describing it metaphorically:

> As an example of a necessary metaphor I may quote the following usages in vogue with peasants when they call a vinebud *gemma*, (a gem) (what other term is there which they could use?), or speak of the *crops being thirsty* or the *fruit suffering*.[32]

This is a familiar phenomenon both in everyday life and in science. As Darwin himself observes:

> [W]hoever objected to chemists speaking of the elective affinities of the various elements? – and yet an acid cannot strictly be said to elect the base with which it in preference combines. It has been said that I speak of natural selection as an active power or Deity; but who objects to an author speaking of the attraction of gravity as ruling the movements of the planets? Every one knows what is meant and is implied by such metaphorical expressions; and they are almost necessary for brevity. So again it is difficult to avoid personifying the word Nature.[33]

What will usually happen in such cases is that the word used metaphorically will eventually die and become the standard literal term.

---

[32] Quintilian (ca. 95CE) VIII ch. 6. John Turner pointed out to us that Quintilian has actually put things the wrong way round here: calling a vinebud '*gemma*' is in fact the original meaning of the word, and calling a jewel '*gemma*' the metaphorical extension. This doesn't of course affect Quintilian's point.

[33] Darwin (1872), p. 63.

The examples cited by Quintilian and indeed many of the examples from the history of science are relatively trivial – the giving of a name to a phenomenon that will be already familiar to writer and audience. Darwin, in the two central cases of 'struggle' and 'select', is doing something more ambitious and complex. Firstly, in asking us to see 'a plant on the edge of a desert' as 'struggling for life against the drought', he is inviting us to look at the phenomena in a radically novel way, in seeing the relation of the plant to the desert as analogous to the relation of a dog to another dog, fighting over a bone. He is grouping together phenomena in a highly unusual way, a way that indeed only makes sense within the context of the overall theory that he is developing in the *Origin*.

But secondly, whereas in Quintilian's examples we were concerned with defining a common property possessed by its instances, in the case of the two most basic concepts of the *Origin*: *struggle* and *selection*, the objects fall under these concepts not because they possess a common intrinsic property, but because they can be seen as related by analogy. These concepts are different from but analogically related to the everyday concepts of *struggle* and *selection*. We may look at the case of *struggle*, where Darwin is most explicit about his procedure.[34] What the reader is to grasp from Darwin's text is a concept that is applicable to an extraordinarily diverse range of phenomena – to perceive in what way the examples given are analogically related ('This canine is to that canine as this plant in the desert is to the drought' etc.).

To understand Darwin's procedure at this point, we may look at a passage from Aristotle, discussing a similar case:

> In fact, 'actuality' means the presence of the thing, but not in the way we call 'potentially', e.g., when we talk of 'a (statue of) Hermes in the wood', and 'the half-line in the whole', i.e., it could be separated out; in the same way we call a man 'a student' even while he is not studying, if he is capable of studying. That which is present in the opposite sense to this is present actually. What we mean is clear in each case from an induction of cases, and it is not necessary to seek a definition of everything but to grasp the analogy, that it is as (1) what is building is to what can build, and (2) what is awake is to what is asleep, and (3) what is seeing to what has its eyes shut but has sight, and (4) what has been separated out of the matter to the matter, and (5) what has been worked up to the not thoroughly worked. Let actuality be set down as one side of this division and let potentiality be the other.

---

[34] Darwin (1859), p. 62.

But things are not all said to exist actually in the same way, but by analogy – 'as this is in that (or to that), so this is in that (or to that)' for the relation is either that of movement to potentiality, or of substance to some particular matter.[35]

Here Aristotle is examining the concept of *potentiality*. He illustrates the extraordinary diversity in the cases where we would say that A was potentially B. He then stresses that this diversity means that we do not seek a common property possessed by all these cases, or seek a definition of the concept, but instead grasp the underlying analogy that leads us to group these highly varied cases together.

Darwin proceeds in essentially the same way as Aristotle, but with the significant difference that whereas Aristotle is seeking to analyse an already familiar concept, Darwin is seeking to introduce a new concept to the reader. Because of this, whereas Aristotle can cite everyday examples where we talk of 'potentiality', Darwin has to resort to the possibilities of metaphor to introduce his general concept of *struggle* ('I use the term Struggle for Existence in a large and metaphorical sense'). He first cites a case where it is clearly possible to talk literally of a struggle – two canines fighting over food, and then sketches out in one short paragraph a range of cases which are also described as 'struggles'. These cases are heterogeneous: not only do they range over vegetation as well as animals, but also in some cases, creatures are seen as struggling with other creatures of their own species, in some, with creatures of a different species, and yet again, in some, with features of their environment; but, most significantly, the relations of things to that with which they struggle look remarkably diverse, including cases that look like co-operation, or as Darwin himself says, dependence, as well as clear cases of antagonism.

What Darwin is doing is inviting readers to find a way in which all these examples can be described as struggling – a standpoint from which they can see all these diverse relations as analogous to the two dogs fighting. The reader who is only familiar with the everyday sense of 'struggle' must initially take these descriptions as metaphorical. In this way, the metaphorical use of the word 'struggle' in this paragraph serves the purpose of concept formation – by reading this paragraph, we come to grasp a new concept: *that which can in the appropriate way be metaphorically described as struggling*. Of course, once we have grasped this concept, we can forget about its metaphorical origins, and simply adopt

---

[35] *Meta* 1048a 35–1048b 8.

the word 'struggle' (and similarly 'selection') as part of our scientific vocabulary: it is unlikely that Darwin would himself have been able to give a precise literal definition of this concept, but neither is such a definition necessary. In such a case, the metaphor does its work by eventually committing suicide.

To summarise, are all these uses of metaphor dispensable? Yes and No. Yes, in that both the theory that Darwin is arguing for and the argument for this theory could be stated in plain literal prose. No, in that the metaphors do genuine work in the exposition of the theory.

Conclusion

We may draw together the threads of this chapter by considering how the word 'select', and with it the phrase 'natural selection', are used in the course of the *Origin*. It is clear that the use made of these is an innovation. How are we to understand such innovations? Are the words involved used literally or metaphorically?[36] We do not normally talk of two daisies being engaged in a life and death struggle with each other. When Darwin said that he used the term 'struggle' in talking about a 'struggle for existence' 'in a large and metaphorical sense'[37], he would not necessarily be concerned with niceties of linguistic theory, and this clearly does not settle the matter: there is no reason to suppose that Darwin himself would have been able to give a worked-out answer to such a question. If we look at the phrase 'natural selection', there are prima facie four possible readings, according to which of the two words are used literally and which metaphorically. Surprisingly, each of these four readings make sense, and each can be seen as required at different stages of Darwin's use of the phrase, and of the metaphors that can be regarded as extensions of it. We can, slightly artificially, present these four readings as a sort of progression, leading to a final stage in which the phrase can be simply accepted as the literal name for a particular theory of evolution.

We begin by drawing an analogy between the activity that humans engage in when they select which animals in their stock to favour and the way that the environment increases the probability that animals possessing certain traits will survive and reproduce and decreases the probability that animals possessing certain other traits will do so. Here, 'select' has as its literal sense a deliberate choice made by human beings, or possibly other

---

[36] White (2001, 2015).   [37] Darwin (1859), p. 62.

rational decision-making creatures, and which hence can only be applied metaphorically to natural processes in the wild. This leads to the first possible reading of the phrase 'natural selection' in which the word 'natural' is used literally, but the word 'selection' metaphorically. This is the starting point for Darwin insisting that the processes he is concerned with do not involve nature being regarded as a designing intelligence driving the direction of evolution. Thus, it means 'process that may be compared metaphorically with human selection, but *qua* natural process is not literally one of selection at all'.

Initially, in the metaphorical use of the word 'select', we are envisaging a process in which there is no agent selecting: instead we are talking about the effect brought about by an amorphous set of features of the environment impinging on creatures. But if we are to explore the analogy between artificial and natural selection by extending the initial metaphor, we need to posit an agent, and so by personifying Nature (the sum total of the features of the environment in which creatures live) as a superhuman farmer engaging in a massive breeding programme encompassing the whole animal and vegetable kingdoms. Now the whole phrase 'natural selection' is to be regarded metaphorically: a natural process being regarded metaphorically as Nature selecting. There are a series of metaphors, particularly in chapter IV, where Darwin personifies Nature ('she'): 'It may be said that natural selection is daily and hourly scrutinising, throughout the world, every variation, even the slightest.'[38] These metaphors are best regarded as extensions of an original metaphor in which nature, as a supervisory goddess, selects. Here metaphor provides Darwin with a convenient tool for *exploring* and working out his initial analogy. What could only be said cumbersomely if we restricted ourselves to literal language can be expressed just as precisely but with ease in metaphor.

However, Darwin is not simply extending his initial metaphor in order to explore the question how aspects of human husbandry are replicated in the natural world. One of his main purposes is to contrast human breeding practices with what happens in the wild. Nature not only selects, she does so much more efficiently than humans ever could. That is to say, Darwin is using the word 'selection' in an analogically extended (literal) sense to cover both human and natural processes of discrimination. This is required whenever Darwin is *arguing* from what is true about artificial selection to

[38] Ibid., p. 84.

what is true about natural selection. We may summarise the argument at
the beginning of chapter IV as follows:

> When humans select, they thereby create new varieties.
>
> Nature selects much more efficiently and powerfully than humans.
>
> ∴ It is possible that when Nature selects, she should thereby create new
> species.

The validity of such an argument requires us to give the word 'selection'
one meaning that is constant throughout the argument, and in the second
premise we also require the word 'select' to be given a meaning that is an
analogical extension of its original meaning to denote any form of discriminatory activity, whether artificial or natural. With this, we have to countenance a third possible reading of 'natural selection', in which 'Nature' is
still personified as an agent, and thus used metaphorically, while 'select'
has a new literal sense.

There is one further possibility to consider: the word 'select' as applied
to Nature is used in a special, technical, sense for use in biology, this
sense being an analogical extension of its sense in human applications.
Clearly such a use enables biologists to go about their work without
worrying whether what they say when they talk about selection would be
equally applicable to everyday cases of human selection, and it is in
precisely this way that biologists proceed nowadays. However, this is
only possible once the theory of evolution by natural selection is well
established and understood. It cannot be used in first introducing the
theory. It would therefore be anachronistic to read such a use back into
the text of the *Origin* itself. Towards the end of the *Origin* Darwin
writes:

> When the views entertained in this volume on the origin of species, or
> when analogous views are generally admitted, we can dimly foresee that
> there will be a considerable revolution in natural history . . . . The other
> and more general departments of natural history will rise greatly in
> interest. The terms used by naturalists, of affinity, relationship, commu-
> nity of type, paternity, morphology, adaptive characters, rudimentary and
> aborted organs, &c., will cease to be metaphorical, and will have a plain
> signification.[39]

---

[39] Ibid., pp. 484–485. In subsequent editions, he writes: 'When the views advanced by me in this
volume, and by Mr. Wallace in the Linnean Journal . . .'

   That is to say, Darwin himself places the dispensing with a metaphorical understanding of the terms used in evolutionary theory in the future, once the theory has been fully developed. Obviously, at the same time the advantages that Darwin found in giving a metaphorical use to terms such as 'select' will be lost. However, those advantages are largely only of relevance in the course of first establishing the theory of natural selection. When finally metaphor has done its work it can retire, we may take the phrase 'natural selection' as simply the literal name of a specific theory of evolution.

CHAPTER 7

Rebuttals of the Revisionists

It is time to confront the work of commentators who have interpreted Darwin's argument for natural selection as something other than an analogical argument. There are, in our view, four revisionist interpretations that especially deserve close scrutiny, in part because they have had some influence in the field, and in part because the challenges they raise are intrinsically interesting, so that any attempt to meet them promises to throw light on wider issues. Two we have already mentioned, from the philosophers of science Richard Richards and Peter Gildenhuys. A third is from James Lennox, well known for his studies of Aristotle's biology. The fourth is from D. Graham Burnett, a wide-ranging historian of science.

Each has developed a distinctive line of attack. Lennox accepts that Darwin conformed his theorising to the vera causa ideal but proposes that he fulfilled its adequacy requirement not via analogical linking of artificial and natural selection but via imaginary illustrations of natural selection in action – what Lennox calls 'Darwinian thought experiments'. In Lennox's analysis, criticism of the analogical interpretation is implicit. Richards, Gildenhuys and Burnett criticise that interpretation explicitly, but without speaking as one as to what is wrong with it or what should replace it. As we noted earlier, Richards and Gildenhuys perceive a conflict between contemporary understandings of what made for strong analogical arguments and what Darwin actually wrote. In Darwin's day, they say, analogical arguments were understood to be weakened by contrasts between the things compared, whereas Darwin went out of his way to make bold the contrasts between artificial and natural selection. Burnett diagnoses a different problem. For him, then as now, analogical arguments require the things compared to be discrete and discontinuous, whereas Darwin went out of his way to show that between artificial selection – where the role of directing agency is maximal – and natural selection – where directing agency is wholly absent – there stretches a gap-filling spectrum

of other forms of selection. Beyond their criticisms, each offers an ingenious account of what Darwin's argument for natural selection is, and the
position of the artificial-selection/natural-selection analogy within the
argument. Like Lennox, Richards sees Darwin as fulfilling the vera causa
ideal's adequacy requirement by appeal not to the analogy but to something else: not thought experiments, however, but real ones, accidentally
conducted when selectively bred animals have gone feral. For Richards,
Darwin made use of the analogy merely to help readers get their heads
around how natural selection works. Gildenhuys is more radical, rejecting
the notion that the vera causa ideal guided Darwin's theorising, and even
denying that Darwin recognised analogous kinds of selection. Gildenhuys'
Darwin made use of the analogy between artificial and natural selection,
yes, but only to generalise a selectional process regarded as the same
everywhere, however outwardly different its forms. Burnett's dissent is
different but no less radical. His Darwin appealed to those diverse forms
of selection to fill the gap between artificial and natural selection with a
spectrum so comprehensively continuous that it undermined his analogical
argument even as it was apparently under construction. Far from the
overall effect being a weakening of the argument for natural selection,
however, it is precisely the 'collapsing' quality of this argument from which
it gains its singular strength.

We shall consider each of these revisionisms in turn, starting with
Lennox's. It bears stressing at the outset that, for all the criticism to follow,
we have learned much from each of these four, and we are in their debt for
having made us think much harder than we would have otherwise about
our own interpretation and the grounds for favouring it over others.

Lennox: Darwinian Thought Experiments

A good way into Lennox's analysis is to recall a famous moment in the
'Difficulties on Theory' chapter (VI) of the *Origin*. By way of answering
opponents who doubted whether natural selection could convert a land-
based animal into an aquatic one, Darwin offered the following conjecture,
set out most elaborately in the first edition but retained in all subsequent
editions. Open-mouthed bears, Darwin reported, have been observed
swimming for hours at the insect-festooned surface of a lake. Suppose a
race of bears should continue to feed in that way for thousands of years,
with no interruption in the insect supply or competition from other
would-be feeders. Over the eons, the average bear form would become
ever better adapted to that kind of feeding – the mouth ever larger, the

body ever more aquatic – until, at the end, what had started as a bear would be transformed into something like a whale.[1]

Although Lennox does not mention the bear-to-whale conjecture, it seems a fine example of what he calls a 'Darwinian thought experiment'.[2] According to Lennox, it was by such fictions that Darwin made natural selection persuasive as a process capable of doing all it was credited with – the business, recall, of chapters VI–VIII on the vera causa reading. What is more, for Lennox, noticing this overlooked role for thought experiments helps us understand a feature of the *Origin*'s structure which might otherwise strike us as odd. We expect someone first to argue for a position and then defend it against objections. Yet Darwin gets his defences in first. Why that way around? Because, says Lennox, Darwin aimed, in good logical form, to establish possibility before probability. The abundance of thought experiments is in keeping with the possibility-establishing functioning of the chapters where they are found. And ever since Darwin, when evolutionary biologists have sought to test the explanatory potential of their theories, they too have turned to thought experiments. Here is a positive function for what, in Lennox's view, have been unfairly dismissed as 'just-so stories'.

Lennox is surely right that – to use terminology just being introduced into analytic philosophy of science when he began writing on these topics – how-possibly explanations of the bear-to-whale kind can serve as useful placeholders for the how-actually explanations that we want. Furthermore, how-possibly explanations are an instance of a wider genre, well worth exploring, of fiction-enhanced reasoning in evolutionary biology. Consider that, like Stephen Jay Gould later on, when Darwin wanted to stress the contingent nature of the evolutionary process, he asked his reader to imagine how differently things might have turned out given imaginary changes to the past – bees instead of mammals for our ancestors, or blood vessels a little further away from our eyes. Or consider, against all that contingency, Ronald Fisher's famous line about how the mathematically inclined biologist who wants to know why there are two sexes first of all asks what the consequences would be of having three. Whether or not these imaginary scenarios should all count as 'thought experiments', or all thought experiments be considered as testing explanatory potential (Peter Lipton expressed doubt), Lennox has usefully drawn attention to a fascinating aspect of evolutionary reasoning.[3]

---

[1] Darwin (1859), p. 184.   [2] Lennox (1993).
[3] Radick (2018), p. 139; Fisher (1958), p. ix; Lipton (1993).

But is he right about Darwin looking to fiction in order to show, apropos of the vera causa ideal's adequacy requirement, the power of natural selection? Let us set aside, as Lennox tacitly does, the problem of what to do with all the evidence supporting the analogical alternative. Taken on its own terms, Lennox's reading suffers mainly from the fact that Darwin was an even more active fictionalist than Lennox supposes. The imaginary illustrations in the *Origin* are in no way restricted to the defence of natural selection, in chapters VI–VIII or elsewhere. They crop up everywhere. In chapter I, for example, Darwin at one point asks whether the extreme variability found on farms might be due not to the domestication process, as he thought, but to the initial choosing for domestication of especially variation-prone kinds of plants and animals. He proceeds to answer his question with, among other considerations, a bit of make-believe:

> I cannot doubt that if other animals and plants, equal in number to our domesticated productions, and belonging to equally diverse classes and countries, were taken from a state of nature, and could be made to breed for an equal number of generations under domestication, they would vary on an average as largely as the parent species of our existing domesticated productions have varied.[4]

Or consider the following, from a passage in chapter XIII where Darwin defends not only the grouping together of organisms related by close descent, but the impossibility of selection-driven change leaving in its wake close relatives so dissimilar that a naturalist would wish not to group them together. Here we get thought experiment within thought experiment:

> But it may be asked, what ought we to do, if it could be proved that one species of kangaroo had been produced, by a long course of modification, from a bear? Ought we to rank this one species with bears, and what should we do with the other species? The supposition is of course preposterous; and I might answer by the argumentum ad hominem, and ask what should be done if a perfect kangaroo were seen to come out of the womb of a bear? According to all analogy, it would be ranked with bears; but then assuredly all the other species of the kangaroo family would have to be classed under the bear genus. The whole case is preposterous; for where there has been close descent in common, there will certainly be close resemblance or affinity.[5]

---

[4] Darwin (1859), p. 17.   [5] Ibid., p. 425.

The conjectural flights that interest Lennox show up all over the *Origin*, not just in the 'adequacy' chapters. But, one might reply on Lennox's behalf, isn't that true of analogical reasoning as well? Consider Darwin's already-quoted testimonial: 'Analogy would lead me one step further. . . . But analogy may be a deceitful guide'. That is from the *Origin*'s concluding chapter, in a passage where Darwin raises the question of how many trees of life there are. On the side of there being just one all-embracing tree, he notes that since, as he has shown, the organisms within a hierarchical system of classification should be regarded as descendants of a common ancestral species, it would seem to follow by analogy that any apparently isolated systems now identified should be considered parts of a super system, their members belonging to a single family tree. He went on to say why, though analogy may mislead, it probably did not do so here, given all that animals and plants of the most varied kinds have in common, in their chemical constitution, the laws that govern their growth and development, and so on.[6] Yes, the *Origin* is chock full of analogies, and even analogical reasoning. But nowhere in the book does Darwin present an analogical argument with anything like the comprehensiveness of the one he developed for natural selection as like artificial selection but more powerful.

Before taking leave of Lennox, we need to ask what to make of that overall structural problem that, he claims, finds resolution in the thought-experimental nature of Darwin's case for natural selection's power. It is a pseudo-problem. The supposed anomaly disappears once one appreciates that, for Darwin, chapters IX–XIII were never conceived of as supplying the evidence which transformed natural selection from a possibly existing causal process into a probably existing one. As we saw in our Chapter 4, Darwin by that point in the book had, by his lights, supplied sufficient evidence to render probable both the existence of natural selection and its species-making power. Indeed, for Darwin, what distinguished a theory like natural selection from a mere hypothesis like pangenesis ('hypothesis' was his term for pangenesis) was that, whereas the sole evidence rendering a hypothesis probable was the evidence it explained, the evidence rendering a theory like natural selection probable extended far beyond what it explained. To put the point another way, the only reason for believing in pangenesis was that, if it existed, it explained so much. Natural selection explained a lot, as shown in chapters IX–XIII; but in addition, there was independent evidence, set out in I–VIII, for believing that natural selection exists and is powerful enough to produce new species from old.[7]

Richards: When Artificial Selection Meets Natural Selection

Lennox largely ignores the farm-nature relationship that bulks so large in conventional glosses of the *Origin*'s first four chapters. By contrast, Richards regards the relationship as central but deeply misunderstood. In brief, he thinks that Darwin valued artificial selection principally for bringing into being unfit creatures whose rapid elimination under natural conditions shows natural selection in action. In that quasi-experimental sense, artificial selection makes natural selection observable – and thus does Darwin fulfil the existence requirement of the vera causa ideal. As for the adequacy requirement, no more is required to fulfil it, in Richards' view, than for Darwin to link together natural variation and the struggle for existence, as he does fully in chapter IV. Both natural variation and the struggle for existence had, by that point in the book, been argued for as real on grounds having little to do with farms. When Darwin now spells out what happens when natural variation meets the struggle for existence, there is no need for any additional argumentation for him to show that this combination is powerful enough to produce new species. Of course, Darwin here and there does gesture toward an analogous process on farms, where the variations induced under domestication mesh with the selectional activities of human breeders. But that, Richards insists, is just to help readers get their minds around a new idea. Such passages are strictly heuristic; Darwin's case for natural selection's existence and power in no way depends on them. Indeed, it could not have depended on them, since, for Darwin, artificial selection differed from natural selection in crucial ways, and differences, as Darwin knew, weakened arguments from analogy.[8]

To take first Richards' construal of how Darwin met the vera causa ideal's adequacy requirement: for Richards, Darwin had only to enunciate what Gould identified as the argumentative core of the theory of natural selection: (i) inheritable variation exists in nature; (ii) there is a struggle for life in nature; (iii) in that struggle, inherited variations tending to adapt individuals better to their conditions will accumulate, leading eventually to the production of new species. Once (i) and (ii) have been shown,

[7] Hull (2009), esp. 184–185; Radick (2017).
[8] R. A. Richards (1997, 2005). See too R. A. Richards (1998, 2014).

(iii) follows, end of story.[9] And indeed, for us now, (i)–(iii) seem irresistible. But in Darwin's day, a key issue for fellow naturalists – as for many creationists in our day – was whether adaptive modifications in nature can proceed without limit, not whether they happen at all. Moreover, even among those who allowed for modification without limit, as Herbert Spencer did, the selective destruction of the less-than-fully-fit individuals was understood as holding kinds of organisms to type, not as shaping new types from old. The creationist Lyell – the reader whose view of the *Origin* Darwin most cared about – saw selective destruction in exactly these terms. In the light of these points of background, one expects to find Darwin in the *Origin* arguing strenuously for the hitherto unappreciated creative power of selective destruction. And so, on the analogical reading of Darwin's argument, one finds, not least in that remarkable expectation-preparing line: 'Natural Selection, as we shall hereafter see, is a power ... as immeasurably superior to man's feeble efforts, as the works of Nature are to those of Art.'[10] Richards gives no alternative gloss of this line. Presumably he would class it as either an instance of Darwin using artificial selection to help his readers understand natural selection, or as an instance of Darwin acknowledging differences between artificial and natural selection. If the former, it is hard to see what feature of natural selection is being illuminated, except for its greater, species-making power – but that, on Richards' analysis, needs only (i)–(iii) to be persuasive, so why Darwin bothered to assert it here, in company with artificial selection, is mysterious. And if Darwin is acknowledging differences between the two kinds of selection, one can say only that the differences to which Darwin refers are all differences which, in his view, explain why natural selection is so much more powerful than artificial selection – again, on Richards' account, a point that was surplus to requirements.

Turning now to what Richards sees as a quasi-experimental argument for the existence of natural selection, it is, he admits, nowhere to be found in the *Origin*. In his view, however, that absence is not the problem it might seem, for, he points out, the *Origin* was but a hurriedly written abstract of the book Darwin really wanted to write. We get a much better idea of that book, or at least the part of it to do with farms and nature, from *The Variation of Animals and Plants Under Domestication*, published

[9] Gould (2002), pp. 125–126. We have lightly modified Gould's version. Although Gould called it a 'syllogistic core', that is a misnomer, above all because a syllogism always has a certain conclusion, whereas the conclusion here is merely probable (albeit highly probable!).
[10] Darwin (1859), p. 61.

only in 1868. There we find, deep in volume 2, in the second of two chapters on selection, a single paragraph on how natural selection places checks on artificial selection, resisting human efforts to 'breed an animal with some serious defect in structure, or in the mutual relation of parts'.[11] After introducing his theme, Darwin in characteristic fashion piles up the examples, ending with a long list of human-bred animals and plants that would never survive in the wild. Along the way he discusses the case that Richards dwells upon most: the Niata cattle of La Plata. Earlier Darwin describes them as 'monstrous'. Here he reports how badly the cattle fare when left to fend for themselves during times of drought, owing to upturned jaws and lip shapes which, selectively engineered by breeders, render the animals unable to graze on the twigs that would otherwise provide sustenance.

Could this really be Darwin's case for natural selection's existence? Darwin nowhere refers to the Niata cattle that way. In his discussion of them, he seems rather to be taking natural selection's existence for granted, concentrating instead on how some otherwise unintelligible limits on the power of artificial selection become intelligible once natural selection is taken into account. Then too, it would be most curious for Darwin to bury his existence case in a not-very-prominent passage in a not-very-prominent book, published only well after he had published the adequacy case in the *Origin*. Richards gives the impression that Darwin wrote the *Origin* so hurriedly that he simply forgot to include the existence case there. Indeed nine months was fast, but, well, not *that* fast; and he seemed, in the months that followed, to have taken quite a lot of care over the proofs. To miss out the foundation of one's argument would be quite a howler. Be that as it may, for Richards, the important point is that Darwin's experimental quasi-deductive argument for natural selection turned the decrease of fitness brought about on farms to evidential advantage. He needed to start the *Origin* on the farm to make that argument (whenever he got around to it), and also to introduce a process whose workings could, for certain purposes, be usefully compared to natural selection.

Richards' handling of text and context in locating Darwin's argument for natural selection finds its sole justification in the view that, whatever that argument is, it cannot be an argument by analogy with artificial selection. But, as have seen, that view is founded on an incomplete understanding of the forms that arguments by analogy can take. Once

[11] Darwin (1868), 2, pp. 226–227.

the tradition of analogy as proportion is in view, there is no difficulty reconciling the text of the *Origin* with philosophically respectable forms of analogical argument. The hunt for a non-analogical argument becomes motiveless. Nor are the results compelling when judged on their intrinsic merits. As noted, Richards' experimental syllogistic reconstruction creates strange puzzles where, on more straightforward interpretations, none exist. Among these puzzles is why Darwin should ever have backed natural selection at all. Richards stresses the doubts that, in the pre-*Origin* era, some thinkers expressed about drawing conclusions from farms to nature when it came to animal and plant modifiability. For Richards, it seems, if Darwin's contemporaries were dubious about something, it must have been off limits, since he was out to persuade them, and he would surely not have made that job any harder than necessary. But if that were so, the struggle for existence – the source of natural selection's power – should also have been off limits for Darwin, since, again, there were doubts as to how much modification could accrue due to the 'continual strife' (as Lyell put it) of animals and plants in a state of nature. Why should Darwin have been put off in the one case but not in the other? Another question to add to the ones heaped up besides Richards' reconstruction.[12]

Gildenhuys: Selection Is One

It is tempting, at this point, to pass over Gildehuys' reconstruction, since it is premised on the correctness of Richards' claim that Darwin's argument for natural selection has to be non-analogical. But that would be a disservice. Gildenhuys goes his own way in construing the non-analogical argument, and in a manner that is, on the face of it, far more respectful of textual and contextual niceties.[13] The key for him is the astronomer John Herschel's famous *A Preliminary Discourse on the Study of Natural Philosophy* (1830). This book is widely credited with helping Darwin appreciate – in conjunction with Lyell's *Principles of Geology* – the virtues of vera causa reasoning. But in Herschel's book Gildenhuys identifies two other methodological principles which, in his view, better match Darwin's reasoning in the *Origin* than the vera causa principle. The first alternative principle is what Gildenhuys calls 'causal decomposition'. The second

---

[12] For criticisms of Richards' position complementary with our own, see Sullivan-Clarke (2013). For an extended defence of Darwin's use of analogy in constructing his theory, see Milman and Smith (1997).
[13] Gildenhuys (2004).

principle involves analogy, but as a means for discovery, not for justification.[14]

By causal decomposition, Gildenhuys means the subtraction, one by one, of all the known causal influences on a phenomenon. Whatever is left will represent the effect of still-to-be-discovered causes. This principle is most clearly enunciated in the following passage in the *Preliminary Discourse*:

> Complicated phenomena, in which several causes concurring, opposing, or quite independent of each other, operate at once, so as to produce a compound effect, may be simplified by subducting the effect of all the known causes, as well as the nature of the case permits, either by deductive reasoning or by appeal to experience, and thus leaving, as it were, a residual phenomenon to be explained. It is by this process, in fact, that science, in its most advanced state, is chiefly promoted. Most of the phenomena which nature presents are very complicated; and when the effects of all known causes are estimated with exactness, and subducted, the residual facts are constantly appearing in the form of phenomena altogether new, and leading to the most important conclusions.[15]

Here, according to Gildenhuys, is Darwin's method in chapter I of the *Origin*. Seeking to account for the causal influences that make domesticated animal and plant varieties as they are, he goes through a list of not very important causes – the effects of climate, the inheritance of the effects of use and disuse, the correlation of parts due to the laws of growth – until he is left with a residuum (quite large) that these cannot explain. A new cause is needed. That cause is selection.

With selection now introduced, Darwin goes on, says Gildenhuys, to generalise it, in accordance with the second Herschelian principle of generalisation-by-analogy. From selection on the farm, Darwin proceeds, by way of analogy, to selection in nature. That may sound no different from the analogical gloss that Gildenhuys repudiates; but there is a difference. On the analogical gloss, Darwin argues for the existence of a process in nature that is like selection on the farm. On the generalisation gloss, however, Darwin argues that the same process of selection occurs in nature as on the farm. So where, in the former case, there are two separate processes, analogically related, in the latter case, there is just one process, whose multiple manifestations are, however, discoverable via analogical thinking. Fussy though it seems, the distinction is important for

---

[14] On the general matter of Darwin's debts to Herschel, see Pence (2018).
[15] Herschel (1830), p. 156.

Gildenhuys, for two reasons. First, he takes it to make sense of references in the *Origin* to 'the principle of selection', singular. Second, he takes it to make sense of Darwin's explicit cataloguing of the ways in which artificial selection contrasts with natural selection – a no-no, of course, on Richards' account of how analogical arguments work, but, according to Gildenhuys, unproblematic on Herschel's account of how generalisation-by-analogy works. Someone investigating the effects of a given cause should expect, Herschel counselled, for there to be all kinds of differences between the newly discovered effects and the familiar ones. What matters is that underneath all the difference they have one thing in common, namely, their being effects of *that* cause. And in helping one to discover the deeper, causal unity beneath surface diversity, analogical reasoning has a major role to play.

But was that really Herschel's advice? It's easy enough to see how analogy could serve discovery in enlarging the class of a cause's known effects where the unknown effects are like the known ones. But where the unknown effects are quite unlike the known ones, reliance on analogy seems, on the face of it, a disastrous strategy. If one turns to Herschel's pages for illumination, it comes, though not in a form favourable to Gildenhuys. What Herschel advised was rather different from what Gildenhuys reports. On a page he cites, we find Herschel exhorting investigators to follow the example of the great Newton in using analogical reasoning, not to identify new effects of a known cause, but to discover new causes:

> Here, then, we see the great importance of possessing a stock of analogous instances or phenomena which class themselves with that under consideration, the explanation of one among which may naturally be expected to lead to the rest. If the analogy of two phenomena be very close and striking, while, at the same time, the cause of one is very obvious, it becomes scarcely possible to refuse to admit the action of an analogous cause in the other, though not so obvious in itself[.] For instance, when we see a stone whirled round in a sling, describing a circular orbit round the hand, keeping the string stretched, and flying away the moment it breaks, we never hesitate to regard it as retained in its orbit by the tension of the string, that is by a force directed to the centre; for we feel that we do really exert such a force. We have here the direct perception of the cause. When, therefore, we see a great body like the moon circulating around the earth and not flying off, we cannot help believing it to be prevented from so doing, not indeed by a material tie, but by that which operates in the other case through the intermedium of the string, – a force directed constantly to the centre. It is thus that we are continually acquiring a knowledge of the existence of

causes acting under circumstances of such concealment as effectually to prevent their direct discovery.[16]

Herschel's point is not, of course, that the same cause keeps the stone and the moon in their respective orbits, but that the distinctive cause of the moon's orbit becomes discoverable after one notices how like a circling stone is the orbiting moon. What cause stands to the moon as the string-holding hand stands to the stone? That, for Herschel, is an example of the power of analogical reasoning. The passage doesn't remotely support Gildenhuys' generalisation-of-a-cause of Darwin's argument.

A return to Herschel's text thus casts doubt on what Gildenhuys identifies as the second step in Darwin's Herschelian method. A return to Darwin's text casts doubt on the supposed first step. Darwin doesn't represent chapter I overall as an exercise in causal decomposition – a gradual stripping away of the causes of adaptive divergence under domestication until only the selection-produced residuum is left. He introduces the problem of explaining such divergence only *after* he switches from a focus on variation (first half) to a focus on selection (second half). When he introduces selection, he does, it's true, engage in a little causal decomposition. 'Some little effect may, perhaps, be attributed to the direct action of the external conditions of life, and some little to habit; but he would be a bold man who would account by such agencies for the differences of a dray and race horse, a greyhound and bloodhound, a carrier and tumbler pigeon.' Further examples of remarkable adaptive divergence follow, then a reaffirmation that 'we must . . . look further than to mere variability', then finally: 'the key is man's power of accumulative selection'. But all of this happens in a single paragraph in the middle of a chapter that, until then, has stuck to topics relating to the variability induced on farms. The chapter as a whole does not read as a mystery whose solution is selection. What is more, Darwin is far more concerned to impress on the reader the contrasts than the commonalities between selection and the previously surveyed causes of variation. There will be exceptions, and he will always allow for these; but generally he represents selection as different in kind from the other processes. Direct action of the environment, habit and so on merely induce variation in individual organisms or influence the patterns it takes. Selection, however, shapes that variation in adaptive directions (without, note, itself inducing variation in individuals or, when variation is induced, influencing its patterns).

---

[16] Ibid., p. 149. For further discussion, see our Chapter 3.

Near the end of the paragraph we have been discussing, we read: 'nature gives successive variations; man adds them up in certain directions useful to him'.[17]

A last look at a work that Richards highlights, *The Variation of Animals and Plants under Domestication*, will underscore the difference for Darwin between selection as a variation shaper and the other variation-related processes. It will also give us one more occasion to admire Darwin's skills as a systematic analogiser:

> Let an architect be compelled to build an edifice with uncut stones, fallen from a precipice. The shape of each fragment may be called accidental; yet the shape of each has been determined by the force of gravity, the nature of the rock, and the slope of the precipice, – events and circumstances, all of which depend on natural laws; but there is no relation between these laws and the purpose for which each fragment is used by the builder. In the same manner the variations of each creature are determined by fixed and immutable laws; but these bear no relation to the living structure which is slowly built up through the power of selection, whether this be natural or artificial selection.
>
> If our architect succeeded in rearing a noble edifice, using the rough wedge-shaped fragments for the arches, the longer stones for the lintels, and so forth, we should admire his skill even in a higher degree than if he had used stones shaped for the purpose. So it is with selection, whether applied by man or by nature; for though variability is indispensably necessary, yet, when we look at some highly complex and excellently adapted organism, variability sinks to a quite subordinate position in importance in comparison with selection, in the same manner as the shape of each fragment used by our supposed architect is unimportant in comparison with his skill.[18]

Here also is an occasion to address Darwin's sometimes writing of selection as if there was just one kind (e.g., 'the power of selection') and sometimes as if there was more than one kind (e.g., 'whether this be natural or artificial selection'). It all depends on the expository job at hand. When he wants to bring out generic features of the selective process, we read of 'selection'. And when he wants to bring out distinctive features of the different kinds of selection, we read of 'natural selection', 'artificial selection', 'sexual selection' and so on. That is all. For Darwin, the cause of

---

[17] Darwin (1859), pp. 29–30.
[18] Darwin (1868), 2, pp. 248–249. On Darwin's architect analogy, as it has become known, see Noguera-Solano (2013); Beatty (2014). We discuss this analogy further in Chapter 8.

new varieties on the farm and the cause of new species in nature were not the same cause, but the analogy was close enough to merit giving the latter a name recalling the former.

These criticisms notwithstanding, there remains, for the determined selection-is-one reader of the *Origin*, a potential escape route. Consider that, on a widely disseminated understanding of the Darwinian tradition, the most important book on natural selection after the *Origin* was Fisher's *The Genetical Theory of Natural Selection* (1930). Fisher's book really was an all-selection-is-one book. For Fisher, if a variant character confers a selective advantage in a particular environment, then whether that environment is characterised by preference-harbouring predators (natural selection), or by preference-harbouring farmers (artificial selection), or by preference-harbouring females (sexual selection) is irrelevant. Anyone who admires Fisher's achievement may feel the tug of a temptation to read Darwin as incipiently Fisherian. Fisher himself encouraged such readings, as in his influential – and deeply misleading – discussion of Darwin and heredity, where he portrayed Darwin as desperate for just the theory of heredity that Fisher favoured.[19] Retrospectively, one can, if so inclined, likewise see Darwin as struggling towards that same summit of abstraction which enabled Fisher's unified description of selection, and accordingly make allowances for the nineteenth-century conceptions – e.g., that humans lie outside of nature – which, on this interpretation, kept Darwin from realising that goal. Note, however, that to embrace teleological history so brazenly would, aside from its general defects, be wholly out of keeping with the analysis that made the all-selection-is-one reading seem plausible in the first place. In common with the other analyses discussed in this chapter as well as with our own analysis, Gildenhuys' is an attempt to read the *Origin* in the light of what came before Darwin rather than what came after him. If Gildenhuys' conclusion can be saved only by jettisoning what is most salutary in his reasoning for it, that seems to us a very steep price.

[19] For an apposite instance of that influence in action, see the famous 1971 essay 'Darwin's Metaphor: Does Nature Select?', by the historian of science Robert M. Young. On Young's account, Darwin's analogical linking of artificial and natural selection, and the metaphors he spun from it, served to paper over a major crack in the foundations of his theory: the absence of a mechanism that could preserve, à la modern genetics, the heritable variations which natural selection was supposed to be powerful enough to accumulate. See Young (1971), especially pp. 98 and 113. For criticism of Fisher's historiography, see Gregory Radick, *Disputed Inheritance: The Battle over Mendel and the Future of Biology*, ch. 1 (forthcoming).

Burnett: Selection Is Spectral

Richards and Gildenhuys share a philosophical conviction about what analogical arguments must be, and on its basis reach a historical conclusion about what Darwin's argument for natural selection cannot have been (namely, analogical). Burnett is largely likeminded, but embeds his reading of the *Origin* within a distinctive big-picture perspective on what analogies do for the advance of scientific knowledge. Over the long run, according to Burnett, the best analogies have been disappearing acts. What they initially and illuminatingly brought together in conceptual union eventually became understood not merely as similar but, in some important sense, as the same. The sublunary sphere was thought of as *like* the superlunary sphere until, between Galileo and Newton, a single set of principles unified them. Clocks were a model for the workings of animate and inanimate nature until, with the triumph of mechanism over vitalism, clockwork became what earthly organisms and heavenly systems *are*, not just what they are like. So too with Darwinian science. It taught people to see selective breeding of new varieties as like what happens in nature, while at the same time – and more profoundly – demolishing that teaching by revealing humans to be fully a part of nature.[20]

Burnett's exegesis of the first four chapters of the *Origin* aims to show how Darwin accomplished this complex trick. For Burnett, these chapters depict artificial selection and natural selection as occupying opposite poles on a continuous spectrum. At the artificial-selection extreme, human agency guides the selection process in a deliberately reasoned way. Here, the posh likes of Sir John Sebright practise what, as we have seen, Darwin calls 'Methodical Selection'. At the natural-selection extreme, agency, human or otherwise, is absent, with selection instead being the upshot of the struggle for existence. And in between is what Darwin calls 'Unconscious Selection': the kind of thing that, as discussed earlier, even the most savage humans will end up doing, simply by favouring certain of their quasi-domesticated animals and plants over others. Filling in the gap on the one side, between Sebright-style self-conscious selection and savage-style unconscious selection, will be a continuum of selective practices, roughly tracking the scale of civilisation. Filling in the gap on the other side, between savage-style unconscious selection and natural selection, will be all manner of interactions between the two, as when dogs kept by

[20] Burnett (2009).

savages have to spend some of their lives fending for themselves in the wild.

In Burnett's view, whatever is linked by continuous gradation cannot serve as terms in an analogical argument, since analogies bring together the discrete and discontinuous. The relationship of artificial selection to natural selection, he tells us, 'is, in fact, not a true analogy at all, if by "analogy" one means a systematic and revelatory juxtaposition of two discrete and discontinuous entities'. The relationship is not one of analogy but spectrum: 'Rather, what Darwin offers is something more like a spectrum – a broad, continuous array of slightly modified instances of a single entity.'[21] That last phrase sounds a bit Gildenhuys-ish; but where Gildenhuys regards Darwin as rapidly establishing natural selection at the most general level, and only then bringing it back down to earth in its various forms, Burnett's Darwin gradually dials down the agency from the Sebright extreme of methodicalness, only reaching natural selection after passing successively through increasingly unselfconscious, increasingly natural-selection-assisted forms of human-guided selection.

Burnett's analysis is at once a defence and a denial of the view that Darwin's argument for natural selection depends on the analogy with artificial selection. It is a defence in the sense that Burnett allows that Darwin puts the analogy to work to make the case for why natural selection should be expected to be more powerful in its effects than artificial selection, classifying the argument overall as an analogical one, albeit of a hitherto unidentified kind: a 'collapsing analogy', falling down even as it goes up. As Burnett puts it colourfully, 'Darwin's striking analogy between natural and artificial selection looks less like a garden-variety analogy than an exotic chameleon capable of quick changes of appearance: here it seems to work as an instructive juxtaposition between two discrete kinds of selection, but moments later it has resolved into a full-spectrum array of minute variations of a single concept of selection.' But the element of denial of a standard way of reading Darwin's analogy is undeniable. At one point Burnett calls it an 'apparent analogy'. Elsewhere he declares it 'slippery': grab hold of artificial selection, and you end up sliding on down to natural selection. In a footnote, he signals broad agreement with Richard Richards.[22]

Although it is the most recent of the revisionisms in our survey, Burnett's shows signs of becoming the most influential; see, for example,

---

[21] Ibid., p. 121.   [22] Ibid., pp. 120, 122, 124, 125 note 8.

the endorsement in Evelleen Richards' magisterial *Darwin and the Making of Sexual Selection* (2017).[23] What, then, is the case against?

First and most importantly, the selection-is-spectral reading starts out from a definition of analogical argument – as the systematic comparison of discrete, discontinuous things – which rules out in advance the existence of the argumentative tradition whose history this book has reconstructed: analogy as proportion. Consider, again, a paradigmatic mathematical example, such as '2 is to 4 as 10 is to 20'. The entities compared are discrete and discontinuous, *and* they can be placed on an interlinking spectrum, the number line. Not, then, analogy or spectrum, but analogy and spectrum, or better still, spectrum-enabled analogy. Next consider a causal counterpart to the mathematical argument, e.g., 'as a small river in flood is powerful enough to wash away a small bridge, so a large river in flood is powerful enough to wash away a large bridge'. One makes such a statement to show that the more powerful cause, because the same in kind but different in degree from the less powerful cause, can be expected, a fortiori, to have effects that are the same in kind as the effects of the less powerful cause, just different in degree. *To say that two causes or two effects are 'the same in kind but different in degree' is just to say that each can be placed at the ends of a graded spectrum or continuum.* Between the small and the big river are to be found rivers of every intervening grade; between the small bridge and the big bridge are to be found bridges of every intervening grade. With analogy-as-proportion arguments, then, spectra are part and parcel of the analogising. So to find, as Burnett rightly does, that selection processes of intermediate status intervene between artificial selection and natural selection is to identify a time-honoured symptom of an analogy-as-proportion argument. Presented as a challenge to an analogical construal of Darwin's argument for natural selection from artificial selection, Burnett's selection spectrum is more truly seen as a vindication of it.

There is a second problem, of a sort that we have already encountered among our revisionisms: the dependence on reading against the grain of Darwin's texts. Nowhere does Darwin himself say that between artificial and natural selection there extends a spectrum, which he then spells out selection process by selection process. That job is left to Burnett. Return to the first chapter of the *Origin*, and what we find is that Darwin initially distinguishes methodical selection from unconscious selection in order to make plain that, when it comes to all that domestication-induced variation documented earlier in the chapter, what matters is that human actions

---

[23] E. Richards (2017), pp. 359–361. See too Inkpen (2014).

have selective consequences for the preservation of that variation, not that the minds behind the actions are crystal clear about what they are doing and why. When thinking is crystalline, as in methodical selection, humans select not merely with breeding in mind but with breeding in the service of a particular end in mind. They want tumbler pigeons with even more pronounced tumbling behaviour, so they allow this maximally tumbling male to mate with that maximally tumbling female, boiling up the rest for pies, then repeating. In unconscious selection, by contrast, humans simply favour some individuals over the rest. There are no breeding thoughts on anyone's mind, let alone thoughts about breeding in the service of a particular end. There are just actions consistent with preferences for, e.g., this entertaining-to-watch pigeon. But the effect of those actions is that some individual pigeons get extra food, extra protection and so on. And so these preferred, pampered few will tend to be the ones who survive to sexual maturity. 'Nevertheless I cannot doubt', wrote Darwin, 'that this process, continued during centuries, would improve and modify any breed, in the same way as Bakewell, Collins, &c., by this very same process, only carried on more methodically, did greatly modify, even during their own lifetimes, the forms and qualities of their cattle'.[24]

A third problem is with Burnett's claim that the spectrum linking artificial and natural selection is a spectrum of agency, running from high to low to zero. Although, again, Burnett shows that something along those lines can be constructed from Darwin's text with enough ingenious quotation-plucking, there is room for doubt whether Darwin would have thought about the spectrum in quite that way. For Darwin, the important point was that artificial selection was a less powerful process of the same kind as the more powerful process of natural selection; so that, where the former produces mere varieties, the latter can be expected to produce new species. Let us consider the shift from methodical selection to unconscious selection. For Burnett, that is a shift away from high-agency artificial selection towards zero-agency natural selection, via closer-to-nature, know-ing-not-what-they-do savages. But for Darwin? As we have seen, if he thought of anything as intervening between artificial selection and natural selection, it would have needed to be more powerful than variety-producing artificial selection. Unconscious selection does not remotely fit the bill. Nowhere does Darwin suggest that, whereas methodical selection can produce only new varieties, unconscious selection can go so much further, producing varieties-unto-new-species (although he does allow that, if we seek to understand what makes for enduringly stable new species, unconscious selection is more instructive than methodical selection).[25] The crucial point for Darwin is that, whatever the degree of self-consciousness involved, if humans are doing the selecting, then the good being served will be the human good, and the results will, for that reason, never add up to modifications that transform a variety into a new species. Only nature, selecting for the good of the organism, can transform an existing species into a new one.

It seems to follow, on this reasoning, that to bring Sebright-level, variety-making powers of selection closer to the species-making selective power of nature, agency would need to go *up*, not down. And indeed, as noted earlier, in Darwin's 1844 *Essay*, he accordingly imagines a super-human being, able, thanks to enhanced faculties and foresight, to modify existing varieties far more extensively than any actual human breeder could ever manage:

> Let us now suppose a Being with penetration sufficient to perceive the differences in the outer and innermost organization quite imperceptible to man, and with forethought extending over future centuries to watch with unerring care and select for any object the offspring of an organism produced under the foregoing circumstances; I can see no conceivable reason why he could not form a new race (or several were he to separate the stock of the original organism and work on several islands) adapted to new ends. As we assume his discrimination, and his forethought, and his steadiness of object, to be incomparably greater than those qualities in man, so we may suppose the beauty and complications of the adaptations of the new races and their differences from the original stock to be greater than in the domestic races produced by man's agency.[26]

In a last-ditch-effort spirit, one could try, with Burnett, to rescue the notion of a down-from-Sebright agency spectrum by insisting nevertheless that unconscious selection is equivalent to a weak form of natural selection. But though Burnett interprets passages from the *Origin* along these

---

[25] Again, as stressed in Chapter 5, Darwin held *mass* unconscious selection, where more durable varieties emerge more slowly because of the frequency of interbreeding, to be similar to natural selection when it acts in the midst of frequent interbreeding. On this theme see Alter (2007).

[26] Darwin (1844), p. 114. On this Being in the theorising of the *Sketch* and *Essay*, see our Chapter 3. Although Robert Richards (2009, pp. 57–59) regards Darwin's speculation as revealing the German Romantic roots of his theorising, because showing that natural selection for Darwin was 'an intelligent and moral force', a closer precedent culturally lies with the British tradition of natural theology, where Robert Boyle and others famously depicted God as an artful contriver working through Nature far more skilfully than any human. See Radick (2009, pp. 155–158) and, for further discussion, our Chapter 8.

lines, one of them merely says that, under different conditions of life, the same savage-kept domesticated animals could be expected to diverge into distinctive varieties, due to occasional exposure to natural selection; and the other that, provided all the savages keeping dogs (or whatever) share preferences, then the free interbreeding of those dogs (or whatever) need not undermine all variety improvement.[27] If Darwin has a default position on how to think about what happens when artificial selection meets natural selection, it is probably that emphasised by Richards: they tend to pull in different directions. And indeed, that is, ultimately, why artificial selection will only ever produce new varieties, and why only natural selection can produce new species

A final word on Darwin's analogising and the Burnettian big-picture view of analogy in the history of science. For Darwin, who learned to understand the significance of Newtonian mechanics from the likes of Herschel, the uniting of heavenly and earthly motions by Newton was, as in the passage from Herschel quoted earlier, a triumph of analogical argumentation. And when Darwin reasoned about organisms as machines, he could be extravagantly analogical in his argumentation, as in this famous passage from his 1862 book on orchids:

> [I]f a man were to make a machine for some special purpose, but were to use old wheels, springs, and pulleys, only slightly altered, the whole machine, with all its parts, might be said to be specially contrived for that purpose. Thus throughout nature almost every part of each living being has probably served, in a slightly modified condition, for diverse purposes, and has acted in the living machinery of many ancient and distinct specific forms.[28]

Darwin's generation saw no conflict between analogical reasoning and the march of science. It is later generations, grown suspicious of the former, that have sought to expunge it from the record of the latter.[29]

In sum, for all the insightful provocation of the revisionisms examined here, none has succeeded in showing that Darwin's argument for natural selection is anything other than an analogical argument. To classify that argument correctly is, of course, an intellectual good, and can be helpful, as we have seen, both in understanding the *Origin* and in relating it to one of

[27] Burnett (2009), pp. 122, 123.
[28] Darwin (1862), p. 348. For a stimulating discussion of how Darwin acquired the view of technology expressed in this analogy, see Pancaldi (2019).
[29] On the lives of analogy in the intellectual culture to which Darwin's generation was heir, see Griffiths (2016).

our culture's oldest traditions in the construction of arguments. But there are larger, sometimes surprising, ramifications as well. We turn next, in our concluding chapter, to consider how, on some of the most far-reaching questions about the *Origin* and its place in science and history, an appreciation of the analogical nature of Darwin's argument turns out to open up new possibilities.

CHAPTER 8

Wider Issues Concerning Darwinian Science

Here we rest our case for our view of the structure and function of Darwin's selection analogy – and for our opposition to any revisionist opposition to any such view. What remain to be treated are the broader historiographical, philosophical and socio-economic themes and issues we think our view can clarify. These come in two clusters. The first concerns Darwin's alignments or otherwise with traditions and innovations preceding his writing of the *Origin*, such as the understanding of art–nature relations, and mechanistic versus animistic philosophies of nature. The second concerns the continuity or otherwise between Darwin's theorising about evolution by natural selection and Darwinian theorising since.

Darwin and 'Aristotelian' Traditions

We begin by looking at how Darwin's work relates to the pre-Darwinian, broadly Aristotelian tradition of biology, and how the *Origin of Species* came to supplant Aristotle's teaching as a dominant starting point for biological research. But first of all, we take a brief glance at how Aristotle has figured in our book so far, as the initiator of an understanding of argument by analogy and of metaphor that we have argued has been appropriate for the interpretation of Darwin. We have emphasised that Darwin would have had scant knowledge of any of the writings, apart from Paley's *Natural Theology*, which we have looked at in Chapters 1 and 2: his knowledge of Aristotle in particular seems to have been no more than inaccurate hearsay, at least at the time of his writing the *Origin*.[1] There can therefore be no question of the writings studied in our earliest chapters forming an influence on Darwin, and the point of our including them was not to imply any such influence.

---

[1] Cf. Gotthelf (1999), pp. 3–30.

In any case Darwin did not need any knowledge of the theoretical discussions of argument by analogy to construct a near perfect such argument. People have spontaneously throughout history used argument by analogy, both in its Reidian form and its Aristotelian form. We may look here at one example:

> And the LORD sent Nathan to David. He came to him, and said to him, 'There were two men in a certain city, the one rich and the other poor. 2 The rich man had very many flocks and herds; 3 but the poor man had nothing but one little ewe lamb, which he had bought. And he brought it up, and it grew up with him and with his children; it used to eat of his morsel, and drink from his cup, and lie in his bosom, and it was like a daughter to him. 4 Now there came a traveler to the rich man, and he was unwilling to take one of his own flock or herd to prepare for the wayfarer who had come to him, but he took the poor man's lamb, and prepared it for the man who had come to him. 5 Then David's anger was greatly kindled against the man; and he said to Nathan, 'As the LORD lives, the man who has done this deserves to die; 6 and he shall restore the lamb fourfold, because he did this thing, and because he had no pity', 7 Nathan said to David, 'You are the man'.[2]

David, having made Bathsheba, the wife of Uriah the Hittite, pregnant, engineers the death of Uriah in order to marry Bathsheba. Nathan comes to confront David with what he has done. What he does is something entirely natural: he constructs an analogous case, and invites David to judge it. David, by judging the rich man, condemns himself. What we have here is a completely valid argument by analogy: despite the differences between the two cases, every ground for condemning the rich man is equally applicable to David: it is precisely because it is an abuse of power, say, that it is to be condemned. (In our terms, Mill's material circumstance is satisfied.) It is clear that for all this, Nathan need have had no *theoretical* understanding of the way argument by analogy works. The theoretical understanding is a *subsequent* reflection on what people have always done.

So, in constructing his argument, Darwin needed no guidance from any of the authors we discussed in our first two chapters, and we are not suggesting he did. The function of those chapters was to track the emergence of two completely different accounts of argument by analogy and two different analyses of such arguments. This was to enable us in our subsequent chapters to confront the question as to which of these two arguments was Darwin offering us in the first four chapters of the *Origin*,

---

[2] II Samuel, 12, 1–7.

and to provide us with the tools to analyse his argument. Which argument Darwin was using was to be determined purely by an exegesis of the text of the *Origin*.

Turning now to our main question, we can see Darwin as not so much rejecting as bypassing the Aristotelian tradition in biology.[3] If we look at the major advances in pre-Darwinian biology, time and again they are made by people who were consciously working within an Aristotelian background, developing or modifying his agenda for biology: figures such as Harvey, Linnaeus and Cuvier.[4] Unlike many of their medieval predecessors, they were not uncritically copying either his methods or results. So that, for example, Harvey stressed the importance of something largely absent from Aristotle, the use of experiment in science, including the biological sciences, with such experiments replacing Aristotle's basing his work on natural history, his extensive and careful survey of the phenomena. Nevertheless, we are concerned here with people who took their inspiration and starting point, even if in different ways, in what they believed Aristotle had achieved.

We have seen that Darwin's quest for an explanation of species origins was not one that Aristotle could have addressed. Moreover, if he had done, he would likely have assumed that species are fixed, as his metaphysics seems to rule out any possibility of new species arising from old.[5] Darwin's explanatory challenges were therefore not ones Aristotle could help with.

But at the same time, although the approach to biological questions initiated by Darwin would eventually provide a starting point and dominate much future biological research, there was no rejecting all that had gone before. Instead, what was of value in previous biological research could be incorporated and re-established within the larger picture that Darwin had opened up.[6] For instance, there is apparently considerable

---

[3] For a further discussion of the differences between Aristotle and Darwin, see Hodge and Radick (2009), pp. 248–251.

[4] For a survey and discussion of Aristotle's influence in biology, see Leroi (2015).

[5] The argument hinges on Aristotle's treatment of 'potentiality' in *Metaphysics Iota*. Aristotle had a dynamic conception of the form of an animal – the animal's form governed and gave to the animal the potential to live out its whole natural cycle. But how could you properly ascribe such a potentiality to a neonate in which practically none of that potential was actualised? For Aristotle, actuality was always prior to potentiality, and you could only ascribe a potentiality to something if it had already been actualised: in the case of a neonate, the only way to make sense of such ascription was to see the form as having been inherited from its parents. If so, that seemed to rule out the possibility of an animal having any other form than that of its parents. (Of course, Aristotle knew of defective creatures, but they still possessed the form of their parents, and not a new form: a three-legged cow was still a quadruped.)

[6] Cf. Depew (2021), pp. 261–79.

overlap between Darwinian and Aristotelian biology. In both, a form of adaptationism plays a central role: in the examples where Aristotle is attempting to explain differences between the organs of different animals, he proposes explanations combining how an organ having these characteristics is appropriate for this environment, for this way of life of the animal, with its being adapted to work with the other organs,[7] just as for Darwin it is adaptation to the environment that drives the whole evolutionary process. But this 'agreement' is arrived at by totally different approaches. For Aristotle, it is primarily a matter of empirical observation, leading him to posit as a best explanation for such adaptation a metaphysical principle, 'Nature always works for the best'. Given his commitment to the fixity of species, it is hard to see what else he could have said.[8]

However, for Darwin, nature acts solely in accord with laws that are indifferent to their consequences for living beings. As a result, those consequences are sometimes malign and sometimes benign. Those creatures that were best able to cope with malign environments or to profit from benign environments were then favoured by nature in their struggle to survive and reproduce. Since there is no longer any commitment to the fixity of species, this would lead, as we have seen, to the adaptation of living creatures to their changing environments. As a result, the relation of nature to its products is quite different from what Aristotle thought: for Darwin, nature works in ways that are only *analogous* to purposive human productive activity, whereas Aristotle sees the very same purposiveness in natural and human creativity. For Aristotle it is purposive nature itself that is to be compared to a human artificer, but for Darwin it is only the effects, the achievements, of nature that are comparable to what an artificer produces. Nature is not a pigeon fancier, but acts in ways having the same sort of effects a pigeon fancier has; indeed, although lacking all skill, nature outperforms even the most skilled pigeon fancier.

---

[7] See for example, Aristotle's treatment of the elephant's trunk that we looked at in Chapter 1.

[8] Another example showing the deep difference of approach but at the same time the close outcome of those approaches by comparing Aristotle's contrast between those parts of animals that differ by 'the more and the less' and those parts that are related by analogy, and the Darwinian contrast between common features of parts of animals that are to be explained by common ancestry, and common features that are the effect of natural selection. For Aristotle on nature doing nothing in vain, see Gottlieb and Sober (2017).

Art and Nature

These Darwinian comparisons and contrasts lead directly on to our next question: how did Darwin understand the relations of art and nature? There was an Aristotelian tradition here too. But there was also an oppositional tradition, developed in the seventeenth century among defenders of the new mechanical philosophy, most prominently Robert Boyle.[9] Did Darwin's position align him any more closely with this early modern tradition than with the ancient one?

As we have seen in previous chapters, for Darwin, artful selection by man and art-simulating selection by nature are to be both compared and contrasted. They have very contrasting causes, and yet act comparably on what they both work with: hereditary variation. Darwin draws no distinction between artificial and natural hereditary variation. The causes of such variation in domesticated animals and plants differ only in degree, not in kind, from those same causes acting less powerfully to produce hereditary variation in the wild; so variation on the farm differs in degree but not in kind from variation in the wild.

On and off the farm, hereditary variation for Darwin is, in and by itself, equally artless and powerless to produce cumulative hereditary diversification fitting varieties to man's uses and fancies or to nature's conditions and circumstances. On the farm, artful and, off the farm, art-simulating selectional causation have sufficient powers to do to hereditary variation what that variation cannot do to itself, although, crucially, the struggle for existence off the farm is far more powerful than artful selection on the farm.

If, for Darwin, art as a cause is vastly inferior to nature in the degrees although not in the kinds of its consequences, does it follow that he regarded art generally as owing its origins to an imitation – albeit a feeble imitation – of nature? Not at all. Yes, natural selection caused the first men; but their much more recent descendants did not become selective breeders because they were consciously or unconsciously imitating what they thought nature was doing. Whatever their thoughts about nature's operations, what matters for Darwin is that their actions have been causal-relationally comparable to what he, Darwin, thinks about those operations. Only in that sense have the artificers of greatest interest to him been imitating nature.

[9] Dijksterhuis (1961); McGuire (1972).

Nor is this topic one on which we find evidence of any big changes of mind as Darwin developed his views. As we saw in Chapter 3, in arriving at his selection analogy late in 1838, he was not fulfilling an ambition to learn about the formation of wild species by reflecting on man's making of artificial varieties. On the contrary, for months before that arrival, he was explicitly contrasting those two processes. For Darwin in this period, the instructive comparison was between wild species formations and the forming of natural domesticated varieties, brought about by natural means, not the forming of artificial domesticated varieties, brought about by artful means such as selective breeding.

In a notebook entry at about this time – several months before arriving at his selection analogy – Darwin quoted favourably a dictum he found in a book by the seventeenth-century polymath Thomas Browne: that 'nature is the art of God'.[10] When penning this dictum Browne may or may not have been a convert to the mechanical philosophers' new alternative to Aristotelian natural philosophy. But the dictum's appeal transcended those early modern options. For the very same dictum is in Dante, writing three centuries before Browne. Indeed, its pedigree is more Platonic than Aristotelian, and more mediaeval than early modern, belonging typically with Christian and other Abrahamic theologians' quests for congruence between the Craftsman of the *Timaeus* and the God of *Genesis*.[11] As for Darwin, at the time he copied the dictum into his notebook he was still enough of an Anglican theist to warm to such legacies from such quests, and so to view species in the wild, and natural varieties of domesticated species, as works of God's art wrought not miraculously but through divinely instituted secondary causes and laws. In 1859 he was no longer a Christian but a deist, and so, in accord with the original early modern meaning of that label, not an atheist but a theist, believing in God but not in the Bible as His miraculously attested, revealed word; and he could still align himself with a natural theology of nature's causes and laws as God's art, although he mostly left that alignment implicit in the *Origin*.[12]

When we move from Darwin's notebook theorising of summer and early autumn 1838 to his later Malthusian moments, in the months from September 1838 to March 1839, we find new complications in his understanding of art–nature relations. Recall that his pre-Malthusian theory about extinctions – his intrinsic species mortality theory – was itself grounded in analogical reasoning, from arboricultural lore about tree-

---

[10] D 54e.   [11] Wetherbee and Alexander (2018); Shaw (2014).
[12] Brooke (2009); F. B. Brown (1984); Mossner (1967); Dilley (2012).

grafting arts to the natural extending and ending of pre-human species lives entailed, Darwin thought, by sexual generation. And recall how reading Malthus had led Darwin to give up this theory and return to his commitment, first made in early *Beagle* years, to Lyell's Malthusian theory of extinctions: a theory depending on no such analogies between art and nature. Finally, recall too how, by the end of 1838, as that revived commitment became integrated with a new Malthusian theory of wild species formations, and selective picking came to the fore as the natural means of making new species, Darwin reversed his old way of comparing and contrasting wild species and artificial domesticated varieties. Only now did what happens in the wild become reinterpreted as analogous to what happens on the farm.

Once this late-arriving selection analogy was in place, Darwin's thinking about art–nature relations became settled. There was no further wrestling with it over the rest of his life, just as there were no consequential changes in his arguments to and from the analogy. We see this continuity confirmed most directly by Darwin's appeal, in the years after 1859, to what has become known as his 'architect analogy'.[13] Invented to clarify his views on design in nature, and more generally on theological teleology, this analogical likening of art to nature – quoted in Chapter 7 – features a builder deliberately and skilfully making a noble and commodious building. The builder is working in accord with an elegant and fitting design, yet doing so solely by choosing and arranging together scattered, unshaped broken fragments of stone which have fallen, without any unnatural artful intervention, from a rocky precipice. In the analogy, Darwin compares the variations among these artless stone fragments to the artless hereditary variation used by art-simulating natural selection and by artful selection by man; and so the builder's actions are compared with the actions of the struggle for existence in the wild and of the stockbreeder on the farm. Darwin concludes that, just as we admire the builder's skill, not the variation in the stone fragments, as principally responsible for the impressive building built from them, so we should appreciate that selection is the paramount power in the forming of new species in the wild and in the making of new varieties on the farm.

Darwin emphasises that the stony variation is accidental in relation to the design of the building. Likewise, the hereditary variation that man's and nature's selection works with is accidental in relation to the changes in plant and animal structures wrought by that causation. He grants that, in

---

[13] See, in addition to the references in Chapter 7 note 18, Lennox (2013).

another sense, the stony variation is not accidental, in that physical laws such as gravitation, and particular circumstances such as the slope of the rock face, have determined the resultant variation. And so likewise with the hereditary variation: it is appropriately called accidental because, although we are ignorant of its causal determination, we assume it has been determined, albeit by causation unknown to us.

Darwin honed his builder argument in years of correspondence with his good friend, the American Presbyterian Harvard botanist Asa Gray, and gave it canonical exposition in 1868 in two passages in his treatise *Variation in Animals and Plants under Domestication*.[14] Disagreeing with Gray, Darwin held that, although determined by lawful causes instituted by an omniscient God, the particular hereditary variations worked on by natural selection may not be divinely designed for that purpose; or, at least, they do not need to be, because this selection is art-simulating enough to produce adaptive changes when working with variation unrelated to those changes.

In countering Gray in this way, Darwin was led to invoke views he had first articulated thirty years earlier in his private notebooks. Chance, he had then concluded, is to matter as freewill is to mind: for in both there is, as he saw it, a lack of known determinate causation and so an illusion of indetermination born of this ignorance. This young Darwin was an avowed necessitarian – in later jargon, a determinist – about matter and mind. And he was a determinist about both because, as he confirmed elsewhere in these notebooks, he explicitly identified the mind with the workings of the brain: an identification he never afterwards either denied or reaffirmed in print.[15] In his 1868 book, his public stance was more humble. He admitted that, in letting his builder analogy argument against Gray lead to these issues about human freewill and divine predestination, he had run out of confidence in his modest metaphysician's competence, and that he wished to quit before getting further out of his depth in ending his treatise.

Aristotelian and Boylean Traditions

Darwin's general views on art and nature, as integrated with his selection analogy, allow us to appreciate how little affinity he had with either of those two art–nature reflective traditions: the Aristotelian and Boylean. We

[14] Darwin (1868), 2, pp. 248–249 and 430–432.   [15] C 166, M 19 and M 57.

may start with Aristotle himself.[16] His conception of art–nature relations was grounded in a contrast and a comparison. He contrasts a craftsman-constructed bed made, say, of beechwood, with a live branch of wood from a beech tree. Plant the bed in soil, and it will not grow more beds. If anything grows from the bed at all, it will be more beechwood. So, a bed as an artificial artefact has a distinctive structure, but no inherent tendency for that structure to be perpetuated. That means, in Aristotle's terminology, that the bed has no nature. If, however, one were instead to plant the beech branch – a work of Nature, not art – it would act in accord with its nature (which is its soul) and grow into another beech tree.

Does this Aristotelian contrast between the artificial and the natural bar any learning from art about nature? Ultimately, yes it does. It is true that Aristotle also teaches that nature is like a doctor doctoring himself. For, in Aristotle's view, this doctor is not an agent distinct from the patient on whom he is practising his art; nor, secondly, does he deliberate, any more than a perfectly practised musician does in bringing his art to his instrument. Likewise, for Aristotle, a craftsmanlike acorn works without assistance or deliberation at growing itself into the oak tree that it is potentially. But this twofold comparison does not inform us about how the acorn forms itself in becoming a maturing oak.

Even this briefest sketch of these Aristotelian comparisons and contrasts between art and nature shows that Darwin was not aligned with this tradition. For Darwin, the stockbreeder's mindful art is in not making an individual become actually what it is now potentially. For the breeder is going to work on a whole herd or orchard of individuals in making multiple improved varieties that would not make themselves without selective breeding. Likewise with the struggle for existence in the wild: its artless but art-simulating selective breeding is not making any individual actually what it is now potentially, but is making multiple varieties which can become many species. And nature's efficacy for Darwin is superior to art's, whereas Aristotle's acorn is not superior to his doctor.

Moreover, selection, artificial and natural, makes a single ancestral stock have many diverse descendants: in the wild, many divergent bird species, say, fitted to diverse ways and conditions of life; and this multiplication and diversification of the one ancestral stock into the many diverse descendants is entirely unlike Aristotle's acorn crafting itself into a mature oak tree. For that acorn has only two possible futures: actualisation of its one maturational potential, as that tree with that nature, or death.

[16] Granger (1993).

This contrast here between Darwinian contingency and Aristotelian maturational potentiality fulfilment is fundamental. Recall what Darwin does and does not say about fishes with gills as ancestors of mammals with lungs. The gill slits in embryo mammals today are evidence of their remote descent from fish ancestors. But, for Darwin, that descent was exceptional, due presumably to exceptional circumstantial contingencies. Only one line of fish species has led to mammal descendants; many more lines have not and never will. There is, then, no inherent general tendency for fish to have mammal descendants, comparable to the law-like tendency of all fish-like gill-slitted mammal embryos to mature into less fish-like lunged mammal adults. Moreover, selection, artificial or natural, serves Darwin's causal–explanatory purposes precisely because it does not work, in accord with any Aristotelean view of nature's art, to actualise a single potential maturational destiny in any ancestral stock. Selection works to make the many diverse descendants from the singular ancestral one; and in nature, what those adaptively diverse many will be is conditioned by diverse future circumstantial contingencies, not by a single ancestral maturational necessitation.

Consider now that other tradition in our pair. Although the mechanical philosophy originally owed most to Descartes' writing in the first decades of the seventeenth century, its eventual stance on art–nature relations was articulated most explicitly in the next generation by Boyle. A devout Christian eager, in the wake of the English civil war, to stress the congruence of the new philosophy with consensus religious opinion, Boyle talked God up and nature down.[17] Unlike natural Aristotelian bodies, Boylean corporeal creatures made from mechanical matter are not differentiated by forms determining distinctive essences. For Boyle, there are no such essences as natures. Nor, then, are there natures as powers in any way akin to arts as powers. Instead, God is omnipotent, and man has the powers of his God-given arts. But when God at creation, and ever since, exercises His powers on matter, as in making an animal, a mass of inert stuff is given structures and functions that it could never give its powerless self.

The same happens when a man makes a clock. God is a great maker and man a lesser one, but nature, in its own right, is not a maker greater than man and less than God. Nature is no maker at all. To admire what is made naturally is to admire God in his works; to admire what a man has made is

---

[17] Boyle (1686); Radick (2009).

to admire his God-given art. To admire the world is to admire a mechanical masterpiece like but infinitely superior to the finest man-made clock.

In this philosophy of nature, artful experimentation with man-made mechanisms can teach us about the natural, lawful, mechanical workings of God since creation. A Boylean clockmaker can thus learn about the workings of the solar system by tweaking his timepiece to make it imitate more closely what God has wrought in the heavens. By contrast, the Aristotelian philosophy offers little encouragement to learning about the natural by studying the artful, since the natural actions of a nature-possessing, self-moving and self-growing beech tree must be contrasted not compared with the unnatural motions that are a craftsman's artful, counter-natural means of making its wood into a bed.[18]

This divergence of the Boylean and Aristotelian stances on learning from experiments may seem to reveal a manifest ancestry for Darwin's analogical inferences from what he calls 'an experiment on a gigantic scale': the century or so of man's expert selective breeding of domestic animals and plants.[19] And this thought may seem compellingly confirmed when one recalls the role in Darwin's education of Paley, natural-theological successor to Boyle and to John Ray; and when one reflects on Darwin's explicitly interpreting eyes as telescopes and orchids as machines; and when one recognises his family's intellectual and economic engagement with what their nation was then most renowned for: its material and ideological construction of machinofacturing capitalism, with its world-beating and world-imitating machines, from steam engines to spinning jennies.[20] Might not such machines and their mechanisms be the key that can be fitted and turned to access the intellectual prehistory of Darwin's selection analogising?

John Cornell and Andrew Inkpen have written insightfully on new, early modern, views about art and nature, especially views grounded in the seventeenth-century mechanical philosophy, which they see as ancestral to Darwin's views two centuries later. And Michael Ruse has recently defended similar ancestral claims. The difficulties with any such historiography are insuperable, however. Domesticated or not, a species for Darwin was not a machine assembled from suitably shaped and arranged parts, nor is a variety. Artificial and natural selection are not machines or mechanisms, because they are not concatenations of material components whose

---

[18] Dijksterhuis (1961).   [19] Darwin (1868), *1*, p. 3.
[20] Radick (2009); Cornell (1984); Inkpen (2014); and Ruse's contributions in R. J. Richards and Ruse (2016).

spatial configurations, connections and motions entail the power of their persistent and consistent selective actions to adaptively diversify many descendants from a single ancestral stock. Nature, with its struggles for existence, works not with passive matter but with the active powers of growth and reproduction that generate selectable hereditary variation, and so nature here is no quasi-Arkwright but a quasi-Bakewell.

What is more, of the various dominant forms of English capitalist life – agrarian, financial, commercial, imperial and machinofacturing – the last of these, usually but surely misleadingly called manufacturing, is arguably the least relevant to the socio-economic contexts of Darwin's science.[21] His father and he (with a paternal loan) bought as capital investments three farms in exceptionally fertile and profitable Lincolnshire, and were wealthy enough to do so partly, perhaps largely, because grandfather Erasmus's second wife had been the widow of a Derbyshire landowner. As for Malthus, it was his optimistic theodicy of ancient empires that explicitly inspired Darwin's teleology of populational wedging and adaptive sorting. Writing as a political economist, Malthus upheld the French physiocrats' view of land for food production as the form of capital fundamental for all those others dependent on it; and he defended the protectionist measures provided by the Corn Laws so cherished by owners of farms and resented by investors in factories. Most obviously, Darwin's grounding of his selection-analogical arguments in Malthusian views of population, food and land, domestic and colonial, and in agriculturalists' selective breeding practices of stock and crop improvement, aligned him with some very distinctively British and gentlemanly alliances: southern and eastern alliances between agrarian, financial and imperial forms of capitalist life, rather than alliances among bourgeois captains of industry and commerce in their northern, western and midland strongholds nearer his childhood home.[22] No wonder that – for his entry in *Bagshaw's Directory* – the middle-aged Darwin, owner of that Lincolnshire land and of the fields around his Kentish home, self-identified as a farmer.[23] No wonder that the families of Darwin, his mentor Lyell and Lyell's mentor Hutton all owned farms in three eastern, coastal counties, fertile thanks to the common remote cause of retreating glaciation after the last ice age. No wonder Darwin's plant and animal geography, with its

---

[21] Hodge (2009b).
[22] The City of London was, and is, England's Wall Street – and one reason Darwin chose to live near the metropolis.
[23] Freeman (1978).

invasions and retreats and territories won and lost, is often tacitly alluding to a global British empire, itself metaphorically reconfigured as the grandest of all British landed estates. Geography, including animal and plant distribution and ethnology, was a science of land and empire.

Mechanism and Animism

The mechanical philosophy did not go unopposed in the eighteenth-century Enlightenment. When David Hume's *Dialogues on Natural Religion* appeared posthumously in 1779, his protagonists openly considered whether the universe was more like a made clock or a generated cabbage, choosing neither option like Hume himself.[24] By that century's close, post-Kantian, post-Enlightenment German romantic idealist philosophers, such as the young Friedrich Schelling, were often critical of Cartesian legacies in Enlightenment France. What attracted them instead was an integrating of cosmic neo-Platonic animism with Spinozistic pantheism to construct comprehensive alternatives to any mechanist or machinist views of nature and art–nature relations. In the German lands this new, romantic anti-mechanism entered the natural sciences under the banner of *Naturphilosophie*.[25]

The biographical case for Darwin being, like Coleridge, an English devotee of German Romanticism has been rested on Darwin's admiration and emulation of that exemplary savant and naturalist Alexander von Humboldt, celebrated for his travels, researches and writings.[26] No such case is convincing, however. Certainly Darwin learned many lessons of numerous kinds from Humboldt, scion of a land-wealthy Prussian family and often regarded as the leading geographer of his day, who did indeed have idealist and Romantic sympathies. But this mentor did not try, even by implication, to teach readers of the books Darwin read – nor does Darwin show any signs of learning from him – how to be a good *Naturphilosophe*, properly aligned with Schelling. For one thing, Humboldt was himself among the critics of *Naturphilosophie*.[27] For another, Darwin's own reasonings to and from art–nature relational comparisons were directly at odds with views of art central to the Romantics' most cherished doctrines.

---

[24] Hume (1779). [25] R. J. Richards (2002)

[26] See the case put by Robert J. Richards in his contributions to R. J. Richards and Ruse (2016). Ruse, in his contributions, shows why the case fails.

[27] See, e.g., Humboldt (1846), *1*, p. 63. Many thanks to Xuansong Liu for this reference.

The Romantics were not always unanimous but they were often agreeing in moving on from Enlightenment confrontations between science and religion and in privileging art as what raises humankind above the rest of creation. Art makes us Godlike and not brutish. But this was art as poetic – that is to say, creative – self-expression, not as skilful technical craft: Beethoven and Schiller, not Bakewell and Sebright. And when comparing nature with art, the Romantics turned to musical and geometrical harmonies, ratios and symmetries, in earthly life and landscapes as well as in the heavens above, just as Pythagoreans and Platonists had done for centuries before. *Naturphilosophen* such as Lorenz Oken carried on that tradition. Romantic nature is artistic, sometimes sublimely so, because it is akin to human fine art, not to any useful, practical human craft.[28] Nothing could be further from Darwin's art–nature reflections and so from his work generally. His sustained likening of the expressive-emotional lives of animals and men in his 1872 *Expression of the Emotions in Man and Animals* made for an especially un-Romantic book.

So, Darwin was no more aligned with German Romanticism than with the mechanical philosophy it sought to replace. But the fact that Darwin seemed to feel no need explicitly to choose between those traditions – between cosmic cabbage and cosmic clock – is itself striking and in need of explanation.

As a first step, consider the obvious limitations Darwin set himself in the *Origin* as in its textual ancestors the *Sketch*, the *Essay* and *Natural Selection*. He does not take on the whole cosmos, nor even the whole solar system in either its first formation or its current operations. Indeed the earth's initiation is barely alluded to, and likewise with the origin of life and the first species. In accepting the limitations concerning the solar system and the earth, Darwin was conforming to a consensus between Herschel and Lyell and Lyell's mentors, Playfair and Hutton, that geological science did not go beyond these temporal and spatial boundaries. In offering to explain how older, extinct species have been replaced with newer ones, but declining to explain how the oldest arose on our once-lifeless planet, again Darwin was aligned with Lyellian precedents.[29]

Lyell presented his generalisations about the physical workings of the earth's surface – his account of balanced aqueous levelling and unlevelling igneous agencies – as a revision and perfecting of Hutton's theory or system of the earth. Significantly, however, Lyell departed from Huttonian precedents in two ways. First, Hutton had grounded his system

---

[28] See Gorodeisky (2016).   [29] Playfair (1802); Lyell (1830–1833).

in his Newtonian natural philosophy of matter and forces: attractive and repulsive forces of gravitation and heat. Second, Hutton had compared this terrestrial system sometimes with a machine and sometimes with an organism. But Lyell invoked no natural-philosophical grounding for his system, and compared it with neither machine nor organism. In closing his *Principles of Geology*, he did, however, implicitly echo a natural-theological Huttonian theme. For Hutton, the Divine purpose of the system of earthly causation is to keep the planet's surface fertile just as farmers – like Hutton himself, author of a manuscript treatise on agriculture – work to maintain their land's fertility. As designer of this maintenance system, the deist Hutton's God is therefore more the ultimate farmer than the supreme mechanic. Lyell was a deist and presumptive heir to a large landed estate in Scotland's famously fertile south-eastern plain, only a day or two's ride north of where Hutton himself had farmed. Had Lyell been more extensive and explicit in the theological ending to his treatise, he could well have allied himself overtly with his predecessor's agrarian-providentialist views.[30]

So, Darwin's declining to commit himself when it came to mechanism versus animism in the *Origin* had a Lyellian warrant. And even when, in other writings, Darwin did address origin questions directly and explicitly – in his theory of generation, pangenesis (published in *Variation*); in his account of humankind's descent from primate ancestors (in the *Descent of Man*); in his unpublished conjectures about life's first beginnings – we still find invoked no commitments about cosmic or about general natural-philosophical (matter, motions and forces) questions, let alone connections between these questions and theorising about land or life or mind.[31]

****

Standing back from the details of this discussion, it is evident that a contextual historiography for the understanding of Darwin's selection analogy needs to place it in relation to at least three clusters of distinctions and options: to do with ancient legacies, national traditions and diverse forms of capitalist life.

Ancient legacies: Readers who concentrated on the chancy, accidental causation Darwin held responsible for the hereditary variation worked on by selection, natural or artificial, often sensed threatening echoes of fortuitous concourses of mindless Epicurean atoms, rather than comforting

[30] Dean (1992); Lyell (1830–1833).   [31] Hoquet (2018).

reminders of a providential, ethically instructive, Stoic physical nature. Those who concentrated on the art-like way the struggle for existence gave individuals with advantageous traits a better chance of survival and reproduction could have agreed that Darwin was indeed in descent from Christian and other Abrahamic Platonists in holding nature to be the art of God, even if Darwin's God is not working at all as Plato's Craftsman did in taking the complete array of species, as eternal, transcendent Forms, to be his model in bringing cosmic order to a material chaos. In sum, whether Darwin's selection analogy looks more Epicurean or Platonic depended on where one put the emphasis. But no amount of emphasis relocation could make the analogy look plausibly Stoical or – excepting the analogy's causal-relational structuring – Aristotelian.

National traditions: We have already noted how attention to the selection analogy makes it implausible that Darwin was a German Romantic. But it does not make Darwin look especially English. That is because, in general, Darwin was culturally more Scottish than English, and – partly because of this – more indebted to French rather than German inspiration and instruction.[32] He was predominantly aligned with Scotland and France's Enlightenments, not with Romantic, idealist, post-Enlightenment Germany; but there was nothing essentially Scottish or French about his arguing analogically as he did; no national scientific culture was especially analogy inclined. As a gentlemanly British naturalist, whether English or Scottish, he was naturally au fait with the selective breeding practices and principles of Sebright, Bakewell and others. But recall that it was Darwin's selection analogy that gave him his agronomical view of art–nature relations, rather than the other way around. He had no such view before he had the analogy; and he had this view as soon as he had the analogy, and was working out its assumptions and implications regarding those relations. As for his education, he could become the first Darwinian partly because of his formal and informal studies with secular mentors in Edinburgh, and despite later studying with priestly teachers at an Anglican church staff college, the University of Cambridge.

Forms of capitalist life: Turning for a final time to the capitalism question, we should recall that the Darwin of the *Origin* has often been read as an individualist – and reasonably enough, given his frequent emphasis on the selection, whether by man or by nature, of hereditarily differing individuals. Just as reasonably, it has seemed plausible to suppose

---

[32] Hodge (2013a). For wide-ranging discussions of diverse views about Darwin's debts and alliances, see Richards and Ruse (2016); La Vergata (1985).

that his individualist inclination must go back to his theoretical thinking before he had his selection theory, that this inclination helped him to arrive at his theory, and that – in line with those Scottish affinities – the ultimate source was that influential Scottish Enlightenment theorist of individualist capitalism, Adam Smith.

However, there are insuperable conceptual and exegetical difficulties for this view. In his mid-1837 notebook theorising, as we saw, Darwin was explicitly integrating two doctrines. The first was that sexual generation serves the interests of species, not the interest of their individual members whose lives are shortened by sexual but extended by asexual generation. Sexual generation prolongs any species' life, by allowing it to postpone extinction by varying adaptively in changing conditions as asexual generation does not. The second was that species are themselves explicitly viewed as quasi-individuals that are born and live and then die of old age like individual animals or plants. Here is a two-level individualism-plus-quasi-individualism that is too complicated, too unorthodox and too peculiar to Darwin to be ascribed to any Smithean or other socio-economic-theoretic source. Its roots lie in Darwin's grandfather's views about sexual and asexual generation, and Darwin's development of his own neo-Brocchian species mortality theorising.

Recall too another directly pertinent theme in this pre-natural-selection-analogy period: namely, what it is tempting to call his species selectionism – species as quasi-individuals differ in their ability to survive and to give rise to descendent species. When, late in 1838, Malthusian reflections led Darwin to drop his neo-Brocchian theory of extinctions and return to Lyell's Malthusian extinction theory, Darwin ceased taking species to be such quasi-individuals, instead regarding them as made by natural selection analogous to inter-individual, intra-specific artificial selection. Thus was Darwin led to the inter-individual, intra-specific natural selection he will argue for in the *Origin*. In sum, Darwin's individualism about most selection, by nature as by man, has Malthusian and agronomical sources, rather than Smithean and political-economical ones.

Charles Darwin, Adam Smith and an Invisible Hand

There is nevertheless a striking similarity between the theory of natural selection and Adam Smith's idea of 'an invisible hand'. In both cases we have a phenomenon that shows such a purposive complexity that to the innocent eye it looks as if it must have been by deliberate design or planning, but which on closer examination is to be explained as the unintended consequence of the actions of a group of individuals who had no such plan in mind but were simply concerned with their immediate situation. This similarity has been frequently remarked and discussed.[33] Our concern here is not the question whether in arriving at his theory Darwin was influenced by Smith.[34] What interests us is the way in which 'an invisible hand' producing an unplanned outcome thereby generates an analogical model of a situation in which a group of agents plan that outcome.

Although Adam Smith only uses the phrase 'an invisible hand' three times, it is basic to his whole thinking. It is most famously appealed to in his defence of free trade. But it is an aspect of his thinking that goes far beyond economics. In fact, his first use of the phrase occurs in his discussion of primitive thinking about astronomy, in *The History of Astronomy*, and the next in a way that is close to our present concern in *The Theory of Moral Sentiments*.[35] However, the most famous use is in *The Wealth of Nations*, in his defence of free trade:

> As every individual, therefore, endeavours as much as he can both to employ his capital in the support of domestic industry, and so to direct that industry that its produce may be of the greatest value; every individual necessarily labours to render the annual revenue of the society as great as he can. He generally, indeed, neither intends to promote the public interest, nor knows how much he is promoting it. By preferring the support of domestic to that of foreign industry, he intends only his own security; and by directing that industry in such a manner as its produce may be of the greatest value, he intends only his own gain, and he is in this, as in many other cases, led by an *invisible hand* to promote an end which was no part of his intention. Nor is it always the worse for the society that it was no part of it. By pursuing his own interest he frequently promotes that of the society more effectually than when he really intends to promote it. I have never known much good done by those who affected to trade for the public good. It is an affectation, indeed, not very common among merchants, and very few words need be employed in dissuading them from it.[36]

The general idea here is that there are a wide variety of situations in which there is a complex collaborative organization of benefit to mankind that gives every appearance of having been deliberately designed, and that can, on examination, be better explained as the unintended consequence of the

---

[33] See, e.g., Sober (1993), pp. 188–191.

[34] It is worth noting here that when Darwin does refer to Smith, whether in his notebooks or *The Descent of Man*, it is invariably in connection with the moral psychology of *The Theory of Moral Sentiments*. See Priest (2017).

[35] In Part IV, ch. 1.   [36] Smith (1776), Bk IV, ch. 2.

behaviour of several individuals each acting purely in their own interest. In fact, for Smith, the remorseless nature of an invisible hand of market forces will prove a more efficient producer of that benefit than fallible human beings who intended to bring it about ever could. Consider, for example, the division of labour throughout a country. No individual within a society can, in isolation, supply all that is necessary for them to lead a satisfactory life; for few people make their own clothes, grow their own food, and so on – even apart from their dependence on others for their knowledge of what is happening in the world they inhabit. To consider a trivial case: if a tailor needs paper clips, she does not improvise them for herself, but purchases them from a shop that has acquired them from a manufacturer of paper clips. In turn, she supplies the manager of the paperclip firm with suits. We have here a vast network of people supporting others and being supported by others, making civilised life possible.

How does such a network develop? It is clearly of far greater complexity than could be devised by any government or individual. However, unfulfilled needs create gaps in the market, and people seeking gainful employment seek out such gaps, gradually developing the complex system as the unintended consequence of countless instances of people seeking to make a living adapting themselves to the economic environment in which they find themselves. Each individual is simply acting in their own interest. However the unintended consequence is that the cumulative effect of vast numbers of people all doing so is the complex division of labour throughout the country. Although individuals are all working in their own interest, they are led by the situation in which they find themselves to behave in precisely the same way as they would do if they were engaged with the rest of the citizens in a vast collaborative project designed to generate such a division of labour.

Of course, Smith's enquiries were purely concerned with transactions between human beings and their human environment, but these restrictions are not relevant to the general idea of 'an invisible hand'. Although there is considerable diversity in the examples that come into consideration here, we can summarise the idea that runs through them all. A large number of individuals, acting independently and considering only what is in their own self-interest, find themselves in circumstances where they are led to act in precisely the way that they would act if they were engaged in a collaborative plan to create a complex phenomenon that has every appearance of being something that must have been designed, but is in fact the unintended consequence of all the individual actions taken together.

This is strikingly reminiscent of what happens in the theory of natural selection, despite differences making it an extension rather than an instance of Smith's invisible hand. To return to the main concern of this book: we are examining Darwin's use of analogies between, and his comparing and contrasting, three different forms of selection – the two types of artificial selection, methodical and unconscious, and natural selection.

Of these, methodical selection was the most straightforward, and was the one for which Darwin could expect familiarity on the part of his readers: it was completely uncontroversial that the deliberate policy of only permitting animals and plants with favourable traits to reproduce could lead to future generations of improved stock or eventually new varieties.

Unconscious selection occurred at the stage at which human beings not yet aware of the possibility of selecting in order to breed new varieties would nevertheless select the best animals for their immediate purposes. So someone would pick a dog that seemed a suitable guard dog with no further thought than that such a dog would fulfil his needs. However, as a result of large numbers of people needing guard dogs making the same kind of selections a variety of dog displaying the characteristics most needed in guard dogs would evolve as the unintended consequence of all the individual selections. One of Darwin's main reasons for claiming that unconscious selection was 'more important' for his purposes than method-ical selection was precisely because it did not need people intending to breed fresh varieties for such varieties to emerge. We have here a very clear instance of an invisible hand in operation that would have been recognised by Smith himself as what he had in mind by talking about an invisible hand.

With natural selection we can no longer talk of humans and their interests, instead the agents of change are frequently inanimate, with no interests whatever, or, say, predators. Predators 'select' in a way that is significantly different from the way humans select. When a man selects a guard dog because it is strong, he thereby promotes the emergence of strength in guard dogs. However, when a fox selects the slowest rabbit, his interest is simply in catching a rabbit, and selects the slow rabbit simply because it is the easiest to catch; he does not thereby promote slowness in rabbits, but rabbits that can run fast enough to be difficult to catch. But despite these differences, the fox behaves in precisely the way he would do if the foxes were embarked together in the project of breeding a variety of rabbit that it would be difficult for foxes to catch.[37] Similarly, the frost acts in precisely the same way that a gardener would act who was intent on breeding frost resistant plants by weeding out those plants that were not so resistant.

In this way we can see natural selection as an invisible hand version of artificial selection, generating the analogy between artificial and natural selection that has been the concern of this book. Whether or not the construction of such an analogical model can guarantee the possibility of the generation of the favourable outcome cannot be answered in general, but must be argued through on a case-by-case basis: what we have been arguing throughout this book is that in the particular case of natural selection, Darwin is justified in arguing that new varieties and species can be formed in the wild by a struggle for existence.

After Darwin: Wallace and Galton

Shifting now from pre-*Origin* to post-*Origin* contexts for Darwin's selection analogy, we begin by noting that, surprisingly, no major reviews of the *Origin* dwelled on its arguments by analogy. Indeed, one comprehensive collection of them has no need for the word in its index; so coming up with generalisations about the immediate reception of the analogy is far from easy.[38] However, two men who were close to Darwin – both his junior by a decade or so – can introduce us to instructive complications.

As we have seen, Wallace had no reason to shun the analogy when first meeting it in Darwin's paper of 1858 and book of 1859, or ever thereafter. But might he have changed his mind thereafter? How else are we to interpret his remarks in the preface to his 1889 book *Darwinism*, where we read, 'It has always been considered a weakness in Darwin's work that he based his theory, primarily, on the evidence of variation in domesticated animals and cultivated plants'?[39]

Reading on, however, one soon learns that this sentence has no bearing on the selection analogy. The weakness Wallace was addressing concerned not Darwin's case for the species-making efficacy of natural selection but his case for the existence of the variation selected in natural selection. And, as we have emphasised elsewhere, Darwin held that the same influences which cause hereditary variation in domesticated plants and animals act,

---

[37] Also, humans behave in precisely the way they would behave if they were seeking to breed antibiotic-resistant bacteria.

[38] Hull (1973). 　　[39] Wallace (1889), p. vi.

albeit less powerfully, on wild plants and animals, and so can be inferred to have the same kind, if not the same degree, of effects. Hereditary variation under domestication served, then, as Darwin's main evidence for the existence of such variation in the state of nature; and Wallace never disputed the validity of this reasoning. But between 1859 and 1889, a number of naturalists had observed, recorded and graphically represented variation in wild animals and plants directly, thus establishing its existence and extent independently of studies of domesticated animals. Wallace in his preface is promising to share with his readers this new knowledge, and its strengthening of the case for his and Darwin's claim that natural selection has plenty of variation to work with in forests, prairies, rivers and oceans.

But what about the evidence for the existence and powers of this natural selection? Here Wallace's book is entirely aligned with the *Origin*. According to Wallace, the potential for geometrical rates of populational increase, together with the actual checks on those increases, entails the struggle for existence, which in turn entails the existence of natural selection, indeed necessitates it so indubitably, Wallace insisted, that any other evidencing is unnecessary. And as for the powers of natural selection, again Wallace hews to the Darwinian line in arguing that these can be inferred analogically from the more observable but less powerful workings of man's selection.[40]

As a theorist concerned to explain adaptive evolution, Wallace needed natural selection – and so this Malthusian evidencing of its existence, and this relational-analogical evidencing of its powers – even more than Darwin did. For Darwin had never rejected what was later dubbed the 'Lamarckian' doctrine of the inheritance of acquired characters; while Wallace's reading in pre-Victorian books by James Prichard and William Lawrence seems to have convinced him, from about the mid-1840s on, that this Lamarckian causation was not at work on the farm or in the wild. By 1889 Wallace could report to his readers that August Weismann's theory of the continuity of the germ plasm was now increasingly taken to have ruled out this causation once and for all. Wallace was with Weismann in holding natural selection to be responsible for all evolution, with one exception not made by the German: the evolution of the human species.[41]

Like Wallace, Francis Galton – a cousin of Darwin's – seems never to have been at all Lamarckian about heredity, and was eventually to be deemed, especially in his own eyes, a pre-Weismannian believer in the

---

[40] Ibid., chapters 4 and 5.   [41] Ibid., pp. 437–440

continuity of the germ plasm. But unlike Darwin, Wallace and Weismann, Galton was no gradualist but a pre-Batesonian saltationist in holding descendant species to arise not in gradual, continuous modifications of ancestral species but in discontinuous shifts from one organic stability to another.[42] But Galton is better known today, indeed notorious, for coining the word 'eugenics' and suggesting that humans should be more selective in their breeding of themselves, the better to emulate improvers of domestic animals. What is not widely appreciated is that, for all the inspiration Galton undoubtedly drew from the *Origin*, his eugenic recommendation needed support only from the first chapter, on variation and selection in domesticated animals and cultivated plants. The eugenic measures Galton envisaged were modifications of the human species as it now exists, and not going so far as to lead to a new successor species, an outcome Galton would have thought beyond the power of selective breeding to achieve.

When it came to explaining the origin of any past, present or future species, whether human or otherwise, Galton had no use, then, for Darwin and Wallace's selection analogy, because he had no use for any arguments from the powers of artificial selection to the much greater powers of natural selection, and no engagement, therefore, with issues about art–nature relations. Galton's prospective human eugenic art would be neither continuing nor imitating the discontinuous, saltationary workings of nature, but instead applying to humans the same practices that, in the name of improvement, human breeders had applied to non-human species.

Both Wallace and Galton had socialist sympathies. Galton saw his eugenic views as socialist in giving priority to societal rather than individual interests. In 1889 Wallace, after decades of associations with socialists, declared himself to be one. Not coincidentally, he wrote the next year on 'human selection', arguing that socialism would lead to two good outcomes: first, a gradual reducing of rates of population increase through delayed marriages; second, facilitation of sexual selection that would steadily eliminate the physically imperfect and the socially and morally unfit. Both of these influences would depend on women becoming, with socialism, better educated and able to choose more freely how they wanted to live and with whom they wished to breed.[43]

Unlike Galton, Wallace's socialism had its sources not in eugenical views but in his youthful Owenite enthusiasm and opposition to landed

---

[42] Bowler (2014); Galton (1889).   [43] Wallace (1890).

capitalism, as practised, he might have noted but didn't, by the Darwin family. Wallace's opposition went back to the 1830s, and later owed much to his reading of Herbert Spencer's *Social Statics* around 1853. The view that land ownership for food production was the primary form of capitalist life lies behind Wallace's subsequent Malthusian species theory just as much as it does Darwin's. But this alignment had quite different implications for the two theorists. In Wallace's 1858 essay, he depicted sub-human wild-animal life as individually self-dependent and Malthusianly competitive, and so exactly what he and Spencer insisted was to be entirely contrasted with the best altruistic, cooperative, social and sympathetic life of civilised men today; and even more tomorrow, when, they believed, socialist hopes would be fulfilled far more completely, particularly if land nationalisation ended ownership of land by capitalist investors not working it as their principal livelihood. For Wallace, then, his 1858 essay was a contribution to his political philosophy, because it implied and supported this fundamental contrast – which became explicit in his 1864 paper on natural selection and the descent of man – between social and sympathetic human life and asocial and unsympathetic animal life.[44]

That moral sentiments are typically sympathetic was another theme canonically tracing to Adam Smith, but to his ethical rather than his economic teachings. Darwin, in explicitly reaffirming this Smithean ethical tradition, has that much in common with the Wallace and Spencer of the 1850s. But Darwin never drew the Spencerian contrast between animal and human life that Wallace did implicitly in 1858 and explicitly in 1864, although he admired both these papers as he would not admire Wallace's writings on man a few years later after their author's conversion to spiritualism.[45]

After Darwin: Wright and Fisher

What of the selection analogy in the twentieth century? In lieu of the comprehensive history still to be written, consider, for now, the life of the analogy in the work of two of the century's leading Darwinian theorists. In 1900, the American Sewall Wright and the English Ronald Fisher were not quite teenagers. But within another dozen years or so they were, unlike many biologists then, agreed in accepting and admiring much that they read in Darwin's *Origin*, and alike too in resolving to use novel statistical analyses to integrate its venerable but controversial principles with the no

---

[44] Wallace (1858, 1864).   [45] Kottler (1985).

less controversial principles of Mendelian heredity, as developed since the first extensive studies in 1900 of the Silesian-born monk's papers of the 1860s. Wright and Fisher went on to become the two most original – and in due course most influential – mathematical theorists of population and evolutionary genetics in the fourth decade of the twentieth century.

Their agreements and disagreements can be easily misunderstood, for three reasons. First, after about 1934, and so some four years after their canonical publications (Fisher's 1930 book *The Genetical Theory of Natural Selection*; Wright's 1931 paper 'Evolution in Mendelian Populations'), their disagreements about evolution became notoriously and irreconcilably bad-tempered. But this is irrelevant here, because, for ten years since meeting in 1924, they were allies and friends, amicably discussing where they differed and where not; and they eventually agreed that they had no mathematical differences to resolve, even while remaining at odds over the causes of evolution.[46] Second, the main divergence is often epitomised as 'Wright's adaptive landscapes' versus 'Fisher's fundamental theorem of natural selection'. But this epitome misleads, for the divergence was really over Wright's causally-very-pluralistic shifting balance theory of evolution and Fisher's not-very-causally-pluralistic theory of evolution by Darwinian mass selection. Third, their distinctive positions on causal pluralism meant that their relations with the Darwinian selection analogy differed instructively. Whereas Wright took on the broad challenge of reinterpreting all of evolutionary causation in the new Mendelian light, Fisher stressed that evolution and natural selection were separate topics, and that he was dealing only with the second. For Fisher, the causal theory of evolution just was the theory of natural selection; and that was what he was rethinking in the light of the new Mendelian genetics.

This contrast is important here because, given his concerns, Fisher could simply endorse and support the old Darwinian analogy, since all selection, artificial and natural, now shared a common Mendelian grounding and mathematical analysis. Indeed, Fisher might even be said to have discarded the analogy, since, on the farm and off, selection for him followed from fitness differences. On that conception, the question of whether selection off the farm is powerful enough to take a population through the species barrier is a non-question. Certainly it didn't arise for Fisher: witness his incuriosity about the evolutionary consequences of natural selection over the long run, and his insouciance about turning fecundity into just another variable inherited character, as much under selection's control as any

---

[46] Fisher (1930); Wright (1931); Provine (1986).

other.[47] But Wright, modelling natural evolution afresh on a new Mendelian rethinking of domestic animal breeding, went beyond the old analogy between artificial and natural selection to an extended analogy between natural and artificial evolution. His new theory of natural evolution – the shifting balance theory – was founded on an equally new theory of artificial evolution. Wright asked: given that evolution is cumulative change in the heredities of species, under what statistical-populational conditions is this change most rapid, continual, irreversible, adaptive and progressive, with or without environmental variation or alteration? His answer: when a large population consists of small local subpopulations with only a little interbreeding between them and inbreeding, random drift and selection within them. As one or more subpopulations become home to selectively favoured individuals, with superior gene combinations, those individuals get exported to other subpopulations, thus contributing to the transformation of the whole population, and ultimately the entire species.

Perhaps a decade before publicly giving this answer to this question about natural evolution, Wright had reached the relevant conclusions about optimum animal-breeding strategy. At Harvard's Bussey Institute, dedicated to agricultural and horticultural science, and drawing on his mentor Castle's work on selective breeding of hooded rats as well as his own doctoral work on guinea pigs, Wright became convinced early on that mass selection of individual traits is often efficacious but also not reliably optimal. From breeding experiments conducted later at the US Department of Agriculture, he confirmed that inbreeding leads to a decline in health and fertility, but can also increase random differentiation among lines and fix distinct trait combinations. At around the same time, his studies of animal breeding histories (especially the detailed long-run North Yorkshire and Durham shorthorn cattle breeding records) led him to appreciate the eventual optimum effects for a whole breed when breeders first combined selection and inbreeding within herds, then exported the sires from the superior herds to the rest for cross-breeding. The upshot was a picture in which, on the farm and off, good breeding – and likewise evolution – appeared to be facilitated best by a 'shifting balance' among the causes and effects of inbreeding, outbreeding and selection, both within and between small local subpopulations.[48]

---

[47] See Fisher (1930), especially 43–44, discussed in Radick (2009), 162. On Fisher's concern with *selection*, not evolution, we have learned much from Alex Aylward's Leeds PhD dissertation, now in progress.

[48] Provine (1986); Hodge (1992b, 2011).

As Wright recognised, however, the path from his conclusions about evolution on the farm in the short run to evolution in the wild over eons was far from straightforward. The most serious problem identified in his 1931 paper arose from sexual reproduction. It is marvellously effective at generating new gene combinations, including some better-than-ever ones. But it is no less effective at chopping up those combinations over future years and generations, as combined genes go their separate ways during the meiotic production of gametes. So how can adaptive evolution be possible in higher animals that can only reproduce sexually? Wright's answer invoked inbreeding as countering and slowing meiotic chopping. He pointed out that plant breeders take advantage of the fact that plants do sexual and asexual reproduction, using sexual reproduction to generate plentiful gene combinations, and then asexual reproduction – grafting and the like – to preserve and perpetuate the preferred combinations, and also preventing more sexual reproduction from breaking them up. With inbreeding, breeders of animals bring sexual reproduction as close as it gets to asexual reproduction in plants – something the breeders had long, if unknowingly, been taking advantage of.[49]

In a 1932 paper, Wright faced another imperfection, this time in selection itself. To follow his reasoning, we need to imagine different degrees of adaptive fitness represented as different altitude contours in a mountainous topography through which populations move. In the wild, natural selection takes a population to the nearest adaptive peak, but cannot move it on to a higher one. Only random drift, taking a population down to a valley and so to the foot of such a taller peak, can do this; and so drift is needed to optimise change. Wright concluded that this need – like the need for inbreeding – was met when there were small local subpopulation numbers.[50]

This breeding-strategy modelling of Wright's was complemented with another, very different modelling, which he owed to Fisher. In 1922 Fisher had argued that mathematical population genetics theory could be conducted as the kinetic theory of gases is: by analysing the distribution of gene frequencies in much the way kinetic theory analysed the distribution of gas particle velocities. Wright soon followed Fisher in adopting this analysis in his quantitative theory, with mathematical results eventually agreeing with Fisher's, although achieved through different mathematical means. But those results never made redundant Wright's qualitative

[49] Note the unwitting Darwinian (Erasmus and Charles both) echoes in Wright's resolution.
[50] Wright (1932).

reasonings in his modelling of natural (in the wild) on artificial (on the farm) evolution: modelling which led him to disagree amicably but fundamentally with Fisher's pure Darwinian mass selection theory of evolution, in preferring his own very different shifting balance theory.[51]

Attention to the fate of Darwin's selection analogy thus helps underscore how far Fisher's optimistic view of the nearly perfect causal partnership between Mendelian heredity and Darwinian natural selection had no place in Wright's theorising, given the imperfections Wright thought needed righting with inbreeding and drift. Furthermore, Wright supported his pluralism about the various causes of evolution, and how they work best when in a delicate shifting balance – his rebranding of Spencer's 'moving equilibrium' – among those factors making for heterogeneity and those factors making for homogeneity (another Spencerian theme), by the same kind of analogical modelling that Darwin had used in less pluralist arguments for the powers of natural selection. In doing this agrarian and horticultural modelling, Wright was more like Darwin than Fisher was. In privileging selection causally, however, Fisher was more like Darwin than Wright was, as Wright acknowledged. As for the profound divergences between the two theorists' views of the world, none went deeper than Wright's debts to Spencer, while Fisher drew on Boltzmann in construing natural selection as a uniquely counter-entropic process.[52]

Natural Selection as a Causal–Explanatory Theory Then and Now

Wright's shifting balance theory became widely discussed, if not widely accepted, thanks to the positive and accessible account of it in Theodosius Dobzhansky's 1937 book *Genetics and the Origin of Species*, along with its later editions and sequels over the next thirty years or so. Dobzhansky's book was the single most decisive contribution to what became known in the 1940s as the 'Modern Synthesis' of evolutionary theory, and so to the Anglophone tradition largely dominant right through to the present day, notwithstanding more recent dissenting calls for it to be replaced by various alternatives. The Wright–Fisher controversy persists, and so too does the pertinence of modelling nature's breeding practices on man's.[53]

One might think that, given the mathematical, Mendelian and molecular-genetical innovations and transformations in the early and middle years of the last century, concern with Darwin's analogy must have become irrelevant. But recall again two features of the *Origin*: the structure

[51] Fisher (1922); Provine (1986).   [52] Hodge (2011).   [53] Dobzhansky (1937); Skipper (2002).

and strategy of its one long argument, and the invocation of chance and chances in relating selection to variation. As we saw, the long argument is structured as it is because it's conformed to the Reidian vera causa evidential ideal; meeting that ideal required Darwin to make a case for the adequacy of natural selection, as a cause of unlimited adaptive divergent descent from singular common ancestral species; and in making that case, Darwin presented an analogy between artificial and natural selection structured in accord with the relational-proportional structure of a canonical Aristotelian causal analogy. As for chance and chances, in both artificial and natural selection, the hereditary variation being selected may arise by chance, by accident; and some of these chance hereditary variants, whether in the wild or on the farm, will confer some better chances and some worse chances of survival and reproduction which is no accident at all. In sum, what artificial and natural selection have in common is persistent and consistent non-random, or non-fortuitous, differential survival and reproduction of random or fortuitous hereditary variations.

Consider again, then, those hereditarily variant red and green bugs feeding on green plants and mostly dying when preyed on by colour-sighted birds, and so providing students in our time, no less than in Darwin's, with an exemplary elucidation of what natural selection is. Call that scenario (1), and then consider three additional causal scenarios:

(2) *Natural drift:* Keep everything natural, with nothing different except that the birds are colour blind. Now natural selection (survival of the fitter) is no longer in action, but random drift (survival of the luckier) is, with the colour difference no longer causally probabilistically relevant to survival, nor then to reproductive success in leaving descendants.

(3) *Artificial selection:* Replace the colour-blind birds in (2) with colour-sighted humans who prefer red bugs over green, and so consistently cull green ones in a spacious indoor cage with plentiful greenery and bug nourishment. These people would be practicing artificial selection today just as in Darwinian times.

(4) *Artificial drift:* Replace the discriminating colour-sighted human cullers in (3) with red-green colour-blind humans. With such indiscriminate cullers in action, there would be random drifting of the frequencies of the red and green bugs, just as with the indiscriminate colour-blind bird predators in (2).[54]

---

[54] Do not worry that biologists do not talk of artificial drift – nor indeed, to hark back to Wright, of artificial evolution. No one does. But we are analysing conceptual distinctions, contrasts and comparisons. Not that we are indulging in unhelpful terminological unorthodoxy. Drift, with its

Likewise, in Wright's shifting balance theorising, one can distinguish at least two more pairings: artificial and natural inbreeding; artificial and natural outbreeding. For with each of these pairings there is a humanly caused model and naturally caused target, and there is learning about nature by knowing about art; and there are artificial means conducing to human ends (better beef or more milk) and natural means conducing to non-human ends (adaptation to life in the wild). Artificial inbreeding is ensured by humanly arranged matings, natural inbreeding, like natural drift, by the consequences of small local subpopulation numbers. Artificial outbreeding is ensured by humanly arranged matings, natural outbreeding by wild freedom to roam within and beyond local subpopulation locations.

These artificial domestic, agrarian causal–explanatory modellings of natural wildlife targets are obviously very much in continuous descent from Darwin's analogical theorising. But for a more comprehensive understanding of this continuity, one needs some threefold comparisons and contrasts. As an initial outline consider three theoretical projects current in our time. There is theorising (a) about how selective breeding goes best on the farm (b) about how it goes usually and optimally in the wild, and (c) about any selection, natural or otherwise, as represented in idealised, abstract mathematical models.

The first two projects go all the way back to Darwin obviously, but the third only to Pearson and others around 1900. The three are easily distinguishable in this simple way; but it is much less easy to clarify consensually how any one can be enhancingly integrated with the other two. For example, both the first two may work with assumptions and conclusions about selection in general, whether artificial or natural, with all selection conceived as any causally non-fortuitous, discriminate differential reproduction of hereditary variants. But the selection coefficients introduced in the mathematical models, as coefficients of any and all differential reproduction of hereditary variants, are indifferent to any assumptions or conclusions about discriminate or indiscriminate causes of that differential reproduction.

All sorts of questions can and have been raised over recent decades about how these three projects may or may not be related to one another. For

effects often construed as cumulative sampling error, and so going quicker in small populations than in large, is usually reckoned to be impossible in those infinite populations of interest mostly to mathematically minded post-Darwinian theorists. Experimentally minded inquirers, Dobzhansky among them, have speeded up drift simply by reducing the finite numbers of individuals in genetically diverse fruit-fly populations in the lab, so producing additional artificial increases in drift.

example, Fisher argued nearly a century ago that Darwin's very concept of a struggle for existence has no place in (a nod here to W. S. Gilbert) a modern-mathematical-Mendelian-genetical theory of natural selection. That much is uncontroversial biography. What is still controversial, however, is whether his argument has been or ought to have been compelling in relation to all three of the projects just distinguished. Quite generally, as this example shows, one can ask whether any of these three projects can appropriately decide for the others how they should interpret their history and conduct their philosophy. And, for another example, one can ask how these three projects can learn from Darwin's bypassing in the *Origin* all those issues associated, in our time, with charges that there is something truistically tautological or fallaciously circular about the theory of natural selection. For consider the quartet of questions he raises and answers about natural selection in the course of his one long argument: What is it? Is it? Could it? and Did it? The first is not a factual, empirical question and is answered with a definitional decision as to what are the causally necessary and sufficient conditions for this causal process. The other three are factual, empirical questions and answered as such, with no attempt at any ontological proof deriving those answers from the definitional essence of this causal process. The theory of natural selection is today arguably similar enough to Darwin's version that distinguishing and addressing these questions as he did – including his selection analogy – can still offer bypasses around those issues and charges.[55]

Concluding Reflections

History, we hope to have shown, is an indispensable resource in interpreting Darwin's analogical argument in the *Origin* from artificial to natural selection. From the vantage point of history viewed at the macroscale, Darwin's argument takes its place within a tradition of analogical argument extending back to Aristotle: the analogy-as-proportion tradition. From the vantage point of history viewed at the microscale – at the level of Darwin's own writings, from his post-*Beagle* notebook theorising onward – the argument takes its place within a tradition of causal–

---

[55] For issues about whether the post-Modern Synthesis theory of natural selection is or is not fundamentally different from Darwin's theory in its conception of natural selection as a causal-explanatory process, see Walsh, Ariew and Matthen (2017); Hodge (2013b); also Hodge (1987); Shapiro and Sober (2007); Godfrey-Smith (2009).

explanatory argument extending back to Newton: the vera causa ideal. Taking Darwin's selection analogy seriously as a product of history in this way in turn illuminates a range of wider topics, from where to situate Darwin as a thinker on the relations between 'art' and 'nature' to how his selection analogy mattered for later Darwinian theorising.

The history offered throughout this book is very much a history of the past, not of the present, much less the future. But we have been normative enough to engage implicitly with questions about what may and even should be happening to Darwin's selection analogy in the years to come. Three papers strike us as especially pertinent to these questions. The first, drawn upon earlier, is Wright's largely but not entirely autobiographical paper of 1978 on livestock breeding and theories of evolution. The second, from 2013, is the historian of science Bert Theunissen's article on Darwin's selection analogy; while the third, from 2009, is the evolutionary biologist Ryan Gregory's on artificial selection, domestication and modern lessons from Darwin's enduring analogy.[56]

Theunissen may appear to end his paper very unequivocally: 'Darwin's analogy between the production of domestic varieties and species formation in nature belongs to the past and should not be used by modern teachers and popularisers to explain the workings of evolutionary theory'.[57] However, his main argument for this conclusion leaves room for discussion. He emphasises that Darwin misrepresented the state of animal breeding lore in his own day, for he did not take proper account of the roles given by breeders to inbreeding and crossbreeding as well as selection. If anything, Theunissen suggests, the problem of misrepresentation has got worse over time, as Darwin's analogy is even more out of line with current accounts of these three breeding practices. One may accept this critique of Darwin's analogy while reflecting that no comparable objections can be levelled at Wright, whose comprehensive analogies between domestic animal and wild animal evolution do indeed take into account the roles given by modern breeders to inbreeding and outbreeding as well as selection. Equally, Gregory's survey is comprehensive enough in its coverage of modern breeding practices other than selection to be largely immune to Theunissen's objections to any Darwinian selection analogies. Be that as it may: even if one thinks that, for purposes of teaching students to understand natural selection in a twenty-first-century

[56] Wright (1978); Theunissen (2013); Gregory (2009).   [57] Theunissen (2013), p. 94.

way, any version of the analogy with artificial selection will be more of a hindrance than a help, it does not follow, in our view, that Darwin's version of the analogy should be off the science syllabus. On the contrary, one of the most salutary lessons that students can learn about science is that even the very best of it always belongs to its historical moment. And one of the most effective ways to drive that critical lesson home is to show students how that holds for past science which we rightly venerate today – the Mendel of the 1866 pea hybrids paper, the Darwin of the *Origin*, and so on. It is, in other words, precisely because Darwin's selection analogy is out of date that it should be taught, with its out-of-dateness not over-looked or apologised for but emphasised and elucidated, in the manner that, we hope, our book has made conspicuous.[58]

On Darwin himself as a subject for future studies, there is a very general thesis that can still do with defending. He has often been depicted as – if caricature is allowed – a boyish natural history enthusiast and genius who got lucky on a voyage; and so not as an intellectual engaging with others in the life of the mind in their time and place. This depiction has been widely discredited of late, perhaps most of all because of recent close attention to the writings from the student, travelling and London years. Our own contribution in this book to the confirmation of Darwin's identity as an intellectual is perhaps principally twofold. We have emphasised his agrarian alignments and we have reached back to Greek antiquity in interpretation of those alignments. In doing so we have gone counter to two widespread presumptions. Intellectuals today are not just themselves typically urban but tend to presume that rural life has been too conserva-tive culturally to be conducive to the innovations intellectuals are supposed to be especially good at originating and elaborating. Turning from inno-vations to continuations, it will have surprised some of our readers that we seem to have departed from a standard take on modern science: namely, that it first made itself modern in Newton's era by finally leaving for dead all the legacies from ancient times. We have been careful not to come even close to implying that Darwin's theories should be read as revised versions of classical Greek or Roman teachings. But we have urged that it is no coincidence that various Latinate terms – *vera causa, a fortiori* and *analo-gia* – prove indispensable in understanding why a book written only a

---

[58] For this historicist case for classroom relevance spelled out in relation to Mendel's paper, see Radick (2016), p. 172.

century and a half ago structures its one long argument as it does; for while Darwin's theories were new, his epistemic ideals were not. To end by circling back to a positive note struck in our Preface, one way to keep Darwinian studies lively and of general interest is to recognise that it is because he lived the intellectual life he did that Darwin, his ancestry and his legacies require and reward hyphenated interdisciplinary studies.

References

Note: The writings by Darwin and Wallace referenced below, as well as a number of the secondary sources, can be found at Darwin Online, http://darwin-online .org.uk/, or at Wallace Online, http://wallace-online.org/.

Alter, Stephen G. 2007. 'Darwin's Artificial Selection Analogy and the Generic Character of "Phyletic" Evolution'. *History and Philosophy of the Life Sciences* 29: 57–82.

Appel, Toby A. 1987. *The Cuvier–Geoffroy Debate: French Biology in the Decades before Darwin*. Oxford: Oxford University Press.

Aquinas. 1265–1272. *Summa Theologiae*. Vol. 3, *Knowing and Naming God* (1a. 12–13). Translated by Herbert McCabe O. P. London: Blackfriars, 1964.

Aristotle. *The History of Animals* (HA).
  Metaphysics (Meta).
  Nicomachean Ethics (NE).
  Parts of Animals (PA).
  Poetics (Poe).
  Posterior Analytics (PostA).
  Prior Analytics (PriorA).
  Rhetoric (Rhet).
  Topics (Top).

Barrett, Paul H. et al., eds. 1987. *Charles Darwin's Notebooks, 1836–1844: Geology, Transmutation of Species, Metaphysical Enquiries*. London: British Museum (Natural History) and Cambridge University Press.

Bartha, Paul. 2019. 'Analogy and Analogical Reasoning'. In *The Stanford Encyclopedia of Philosophy*, Spring 2019 edition. Edited by Edward N. Zalta. https://plato.stanford.edu/archives/spr2019/entries/reasoning-analogy/.

Bartov, H. 1977. '*A fortiori* Arguments in the Bible, in Paley's Writings, and in the "Origin of Species"'. *Janus* 64: 131–145.

Beatty, John. 2013. 'Chance and Design'. In Michael Ruse, ed., *The Cambridge Encyclopedia of Darwin and Evolutionary Thought*. New York: Cambridge University Press, pp. 146–151.

2014. 'Darwin's Cyclopean Architect'. In R. P. Thompson and D. Walsh, eds., *Evolutionary Biology: Conceptual, Ethical, and Religious Issues*. Cambridge: Cambridge University Press, pp. 175–192.

Bedall, Barbara G. 1988. 'Wallace's Annotated Copy of Darwin's *Origin of Species*'. *Journal of the History of Biology* 21: 265–289.

Beer, Gillian. 1983. *Darwin's Plots: Evolutionary Narrative in Darwin, George Eliot and Nineteenth-Century Fiction*. London: Routledge & Kegan Paul.

Berkeley, George. 1732. *Alciphron: Or, the Minute Philosopher*. London: J. Tonson.

Bowler, Peter J. 2014. 'Francis Galton's Saltationism and the Ambiguities of Selection'. *Studies in the History and Philosophy of Biological and Biomedical Sciences* 48: 272–279.

Boyle, Robert. 1686. *A Free Inquiry into the Vulgarly Received Notion of Nature*. London: John Taylor.

Brooke, John H. 2009. 'Darwin and Victorian Christianity'. In Jonathan Hodge and Gregory Radick, eds., *The Cambridge Companion to Darwin*, 2nd edition. Cambridge: Cambridge University Press, pp. 197–218.

Brown, F. B. 1984. 'The Evolution of Darwin's Theism'. *Journal of the History of Biology* 19: 1–45.

Brown, William R. 1989. 'Two Traditions of Analogy'. *Informal Logic* 11: 161–172.

Browne, Peter. 1733. *Things Divine and Supernatural Conceived by Analogy with Things Natural and Human*. London: William Innys; Richard Manby.

Burkhardt, F. et al., eds. 1985. *The Correspondence of Charles Darwin*, 30 vols. Cambridge: Cambridge University Press.

Burnett, D. Graham. 2009. 'Savage Selection: Analogy and Elision in *On the Origin of Species*'. *Endeavour* 33: 120–125.

Cajetan (Thomas de Vio). 1498. *The Analogy of Names (de Nominum Analogia)* and *The Concept of Being*. Literally translated and annotated by E. A. Bushinski and Henry J. Koren. Pittsburgh, PA: Duquesne University, 1959.

Copleston, Edward. 1821. *An Enquiry into the Doctrines of Necessity and Predestination*. London: John Murray.

Cornell, John F. 1984. 'Analogy and Technology in Darwin's Vision of Nature'. *Journal of the History of Biology* 17: 305–344.

Costa, J., ed. 2009. *The Annotated Origin: A Facsimile of the First Edition of on the Origin of Species*. Cambridge, MA: Harvard University Press.

Darwin, Charles. 1839. *Journal of Researches into the Geology and Natural History of the Various Countries Visited by H. M. S. Beagle*. London: Henry Colburn.

1842. 'Sketch'. In Gavin de Beer, ed., *Evolution by Natural Selection: Darwin and Wallace*. Cambridge: Cambridge University Press, 1958.

1844. 'Essay'. In Gavin de Beer, ed., *Evolution by Natural Selection: Darwin and Wallace*. Cambridge: Cambridge University Press, 1958.

1845. *Journal of Researches into the Natural History and Geology of the Countries Visited during the Voyage Round the World of H.M.S. Beagle ...*, 2nd edition. London: John Murray.

1859. *On the Origin of Species by Means of Natural Selection, or the Preservation of Favoured Races in the Struggle for Life*. London: John Murray.

1861. *On the Origin of Species by Means of Natural Selection, or the Preservation of Favoured Races in the Struggle for Life*, 3rd edition. London: John Murray.

1862. *On the Various Contrivances by Which British and Foreign Orchids Are Fertilised by Insects, and on the Good Effects of Intercrossing*. London: John Murray.

1868. *The Variation of Animals and Plants under Domestication*, 2 vols. London: John Murray.

1871. *The Descent of Man, or Selection in Relation to Sex*. London: John Murray.

1872. *The Origin of Species by Means of Natural Selection, or the Preservation of Favoured Races in the Struggle for Life*, 6th edition. London: John Murray.

1958. *The Autobiography of Charles Darwin 1809–1882*. Edited by Nora Barlow. London: Collins.

Dean, D. 1992. *James Hutton and the History of Geology*. Ithaca, NY: Cornell University Press.

Depew, David. 2009. 'The Rhetoric of the *Origin of Species*'. In Michael Ruse and Robert J. Richards, eds., *The Cambridge Companion to the 'Origin of Species'*. Cambridge: Cambridge University Press, pp. 237–255.

2021. 'Aristotelian Teleology and Philosophy of Biology in the Darwinian Era'. In S. Connell, ed., *The Cambridge Companion to Aristotle's Biology*. Cambridge: Cambridge University Press, pp. 261–279.

Diderot, D. and d'Alembert, J. 1751–1772. *Encyclopédie, ou dictionnaire raisonné des sciences, des arts et des métiers*. Paris: André le Breton, Michel-Antoine David, Laurent Durand, and Antoine-Claude Briasson

Dijksterhuis, E. J. 1961. *The Mechanization of the World Picture*. New York: Oxford University Press.

Dilley, S. 2012. 'Charles Darwin's Use of Theology in the *Origin of Species*'. *British Journal for the History of Science* 45: 29–56.

Dobzhansky, Theodosius. 1937. *Genetics and the Origin of Species*. New York: Columbia University Press.

Eddy, Matthew D. 2004. 'The Rhetoric and Science of William Paley's *Natural Theology*'. *Literature and Theology* 18: 1–22.

Euclid. 1933. *The Elements*. Ed. Isaac Todhunter. London: J. M. Dent & Sons.

Fahnestock, Jean. 1996. 'Series Reasoning in Scientific Argument: "Incrementum and Gradatio" and the Case of Darwin'. *Rhetoric Society Quarterly* 26: 13–40.

Ferguson, Adam. 1792. *Principles of Moral and Political Science*. London: A. Strahan and T. Cadell; Edinburgh: W. Creech.

Fisher, Ronald A. 1922. 'On the Dominance Ratio'. *Proceedings of the Royal Society of Edinburgh* 42: 321–341.

1930. *The Genetical Theory of Natural Selection*. Oxford: Clarendon Press.

1958. *The Genetical Theory of Natural Selection*, 2nd edition. New York: Dover.

Fodor, Jerry. 2005. Review of Daniel Dennett. *Times Literary Supplement* 29 July.

Freeman, R. B. 1978. *Charles Darwin: A Companion*. Folkestone: Dawson.

Galton, Francis. 1869. *Hereditary Genius*. London: Macmillan.

1889. *Natural Inheritance*. London: Macmillan.

Gentner, Dedre. 1982. 'Are Scientific Analogies Metaphors?' In David S. Miall, ed., *Metaphor, Problems and Perspectives*. Brighton: Harvester Press Ltd.

Gentner, Dedre and Gentner, Donald R. 1983. 'Flowing Waters or Teeming Crowds: Mental Models of Electricity'. In Gentner & Stevens, pp. 99–129.

Gentner, Dedre and Stevens, Albert L., eds. 1983. *Mental Models*. Hillside, NJ and London: Lawrence Erlbaum.

Gildenhuys, Peter. 2004. 'Darwin, Herschel, and the Role of Analogy in Darwin's *Origin*'. *Studies in History and Philosophy of Biological and Biomedical Sciences* 35: 593–611.

Godfrey-Smith, Peter. 2009. *Darwinian Populations and Natural Selection*. Oxford: Oxford University Press.

Gorodeisky, K. 2016. 'Aesthetics: 19th-Century Romantic Aesthetics'. In *Stanford Encyclopedia of Philosophy*. Online.

Gottlieb, P. and Sober, E. 2017. 'Aristotle on "Nature does nothing in vain."' *HOPOS* 7: 246–271.

Gotthelf, Allan. 1999. 'Darwin on Aristotle'. *Journal of the History of Biology* 32: 3–30.

Gould, Stephen Jay. 2002. *The Structure of Evolutionary Theory*. Cambridge, MA: Belknap Press.

Granger, H. 1993. 'Aristotle on the Analogy Between Action and Nature'. *Classical Quarterly* 43: 168–176.

Gregory, T. Ryan. 2009. 'Artificial Selection and Domestication: Modern Lessons from Darwin's Enduring Analogy'. *Evolution: Education and Outreach* 2: 5–27.

Griffiths, Devin. 2016. *The Age of Analogy: Science and Literature between the Darwins*. Baltimore: Johns Hopkins.

Henle, Paul. 1958. 'Metaphor'. In Paul Henle, ed., *Language, Thought and Culture*. Ann Arbor: University of Michigan Press, pp. 173–195.

Herschel, John F. W. 1830. *A Preliminary Discourse on the Study of Natural Philosophy*. London: Longman et al. Reprinted by Routledge/Thoemmes Press, 1996.

Hesse, Mary. 1963. *Models and Analogies in Science*. London, Sheed & Ward.

Hodge, M. J. S. 1977. 'The Structure and Strategy of Darwin's "Long Argument"'. *British Journal for the History of Science* 10: 237–246. Reprinted in Hodge 2008, article VIII.

1987. "Natural Selection as a Causal, Empirical, and Probabilistic Theory." In Lorenz Krüger et al. (eds.), *The Probabilistic Revolution*. 2 vols. Vol. 2: Ideas in the Sciences. Cambridge, MA: MIT Press, pp. 233–270.

1992a. 'Natural Selection: Historical Perspectives'. In Evelyn Fox Keller and Elizabeth Lloyd, eds., *Keywords in Evolutionary Biology*. Cambridge, MA: Harvard University Press, pp. 212–219.

1992b. 'Biology and Philosophy (including Ideology): A Study of Fisher and Wright'. In Sahotra Sarkar, ed., *The Founders of Evolutionary Genetics: A Centenary Reappraisal*. Dordrecht: Kluwer, pp. 231–293. Reprinted in Hodge 2008, article XIII.

2008. *Before and After Darwin: Origins, Species, Cosmogonies, and Ontologies*. Aldershot: Ashgate.

2009a. *Darwin Studies: A Theorist and His Theories in their Contexts*. Aldershot: Ashgate.

2009b. 'Capitalist Contexts for Darwinian Theory: Land, Finance, Industry and Empire'. *Journal of the History of Biology* 42: 399–416.

2011. 'Darwinism after Mendelism: The Case of Sewall Wright's Intellectual Synthesis in His Shifting Balance Theory of Evolution'. *Studies in History and Philosophy of Biological and Biomedical Sciences* 42: 30–39.

2013a. 'Darwin's Book: *On the Origin of Species*'. *Science and Education* 22: 2267–2294.

2013b. 'The Origins of the *Origin*: Darwin's First Thoughts about the Tree of Life and Natural Selection, 1837–1839'. In Michael Ruse, ed., *The Cambridge Encyclopedia of Darwin and Evolutionary Thought*. Cambridge: Cambridge University Press, pp. 64–71.

2016. 'Chance and Chances in Darwin's Early Theorising and in Darwinian Theory Today'. In Grant Ramsey and Charles Pence, eds., *Chance in Evolution*. Chicago: University of Chicago Press, pp. 42–75.

Hodge, Jonathan, and Radick. Gregory. 2009. 'The Place of Darwin's Theories in the Intellectual Long Run'. In Jonathan Hodge and Gregory Radick, eds., *The Cambridge Companion to Darwin*. 2nd edition. Cambridge: Cambridge University Press, pp. 246–273.

Hoquet, Thierry. 2018. *Revisiting the* Origin of Species*: The Other Darwins*. London: Routledge.

Hull, David L., ed. 1973. *Darwin and His Critics*. Cambridge, MA: Harvard University Press.

Hull, David L., 2009. 'Darwin's Science and Victorian Philosophy of Science'. In Jonathan Hodge and Gregory Radick, eds, *The Cambridge Companion to Darwin*, 2nd edition. Cambridge: Cambridge University Press, 2009, pp. 173–196.

Humboldt, Alexander von. 1846. *Cosmos: Sketch of a Physical Description of the Universe*. Trans. Elizabeth Sabine under the superintendence of Edward Sabine. 2 vols. London: Longman, Brown, Green, and Longmans. Reprinted by Cambridge University Press, New York, 2010.

Hume, David. 1779. *Dialogues on Natural Religion*. London: G. Robinson and T. Cadell.

Inkpen, S. Andrew. 2014. '"The Art Itself Is Natural": Darwin, Domestic Varieties and the Scientific Revolution'. *Endeavour* 38: 246–256.

Isocrates. *To Philip*.

Jantzen, Bernard. 2014. *An Introduction to Design Arguments*. Cambridge: Cambridge University Press.

Joseph, H. W. B. 1916. *Introduction to Logic*, 2nd edition. Oxford: Clarendon Press.

Kant, Immanuel. 1787. *Critique of Pure Reason*, 2nd edition. Trans. Norman Kemp Smith, Macmillan: London, 1933.

1783. *Prolegomena to any Future Metaphysics that will be Able to Come Forward as Science*. Trans. Gary Hatfield. In *Theoretical Philosophy after 1781*, pp. 29–169.

1793. *Critique of Judgment*, 2nd edition. Trans. Werner S. Pluhar. Indianapolis: Hackett Publishing Company, 1987.

1798. *Anthropology from a Pragmatic Point of View*. Trans. Louden. In Günther Zöller & Robert B. Louden, eds., *Anthropology, History and Education*. In the Cambridge Edition of the works of Immanuel Kant. Cambridge: Cambridge University Press, 2007.

King, William. 1709. 'Predestination and Foreknowledge Consistent with the Freedom of Man's Will.'

Kirby, John T. 1997. 'Aristotle on Metaphor'. *American Journal of Philology* 118: 517–554.

Kneale, W. C. 1948. *Probability and Induction*. Oxford: Clarendon Press.

Kohn, D et al. 1982. 'New Light on the Foundations of the *Origin of Species*'. *Journal of the History of Biology* 15: 419–442.

Kohn, David. 2009. 'Darwin's Keystone: The Principle of Divergence'. In Michael Ruse and Robert Richards (eds.), *The Cambridge Companion to The "Origin of Species"*. Cambridge: Cambridge University Press, pp. 87–108.

Kottler, Malcolm. 1985. 'Charles Darwin and Alfred Russel Wallace: Two Decades of Debate over Natural Selection'. In D. Kohn (ed.), *The Darwinian Heritage*. Princeton, NJ: Princeton University Press, pp. 367–432.

La Vergata, A. 1985. 'Images of Darwin: A Historiographic Overview'. In D. Kohn, ed., *The Darwinian Heritage*. Princeton: Princeton University Press, pp. 901–973.

Lawrence, William. 1822. *Lectures on Physiology, Zoology and the Natural History of Man*. London: J. Smith.

Leech, Geoffrey. 1969. *A Linguistic Guide to English Poetry*. Harlow: Longmans.

Lennox, James G. 1993. 'Darwinian Thought Experiments: A Function for Just-So Stories'. In Tamara Horowitz and Gerald J. Massey, eds., *Thought Experiments in Science and Philosophy*. Savage, MD: Rowman & Littlefield, pp. 223–245.

2013. 'Darwin and Teleology.' In Michael Ruse, ed., *The Cambridge Encyclopedia of Darwin and Evolutionary Thought*. Cambridge: Cambridge University Press, pp. 152–157.

Leroi, Armand Marie. 2015. *The Lagoon: How Aristotle Invented Science*. New York: Penguin.

Letter to the Hebrews. New Testament, RSV.

Levine, George. 2011. *Darwin the Writer*. Oxford: Oxford University Press.

Lipton, Peter. 1993. 'Review of Horowitz and Massey'. *Thought Experiments in Science and Philosophy Ratio* 6: 82–86.

Lyell, Charles. 1830–1833. *Principles of Geology*. 3 vols. London: John Murray.

McGuire, J. E. 1972. 'Boyle's Conception of Nature'. *Journal of the History of Ideas* 33: 523–542.

Malthus, Thomas Robert. 1798. *An Essay on the Principle of Population, as it Affects the Future Improvement of Society with Remarks on the speculations of Mr. Godwin, M. Condorcet, and Other Writers*. London: J. Johnson.

Mill, John Stuart. 1843. *A System of logic, Ratiocinative and Inductive, being a Connected View of the Principles of Evidence and the Methods of Scientific Investigation*. London: Longmans, Green, and Co.

Milman, A. B. and Smith, C. L. 1997. 'Darwin's Use of Analogical Reasoning in Theory Construction'. *Metaphor and Symbol* 12: 159–187

Milo, Daniel S. 2019. *Good Enough: The Tolerance for Mediocrity in Nature Society*. Cambridge, MA: Harvard University Press.

Mossner, E. C. 1967. 'Deism.' In P. Edwards, ed., *Encyclopedia of Philosophy*. New York: Macmillan.

Nieuwentyt, Bernard. 1721. *The Religious Philosopher*. London: J. Senex.

Noguera-Solano, Ricardo. 2013. 'The Metaphor of the Architect in Darwin: Chance and Free Will'. *Zygon* 48: 859–874.

Nowottny, Winifred. 1962. *The Language Poets Use*. London: Athlone Press.

Ospovat, Dov. 1981. *The Development of Darwin's Theory: Natural History, Natural Theology, and Natural Selection, 1838–1859*. Cambridge: Cambridge University Press.

Owen, G. E. L. 1960. 'Logic and Metaphysics in Some Earlier Works of Aristotle.' In I. Düring and G. E. L. Owen, eds., *Aristotle and Plato in the Mid-Fourth Century*. Göteborg: Almquist and Wiksell, pp. 163–190.

Paley, William. 1794. *A View of the Evidences of Christianity*. London: R. Faulder.
    1802. *Natural Theology*. The edition cited is from Oxford University Press, eds. Matthew D. Eddy, 2006.

Pancaldi, Giuliano. 2019. 'Darwin's Technology of Life.' *Isis* 110: 680–700.

Partridge, D. 2018. 'Darwin's Two Theories, 1844 and 1859.' *Journal of the History of Biology* 51: 563–592.

Pence, Charles. 2018. 'Sir John F.W. Herschel and Charles Darwin: Nineteenth-Century Science and its Methodology'. *HOPOS* 8:108–140.

Plato. *Philebus*.

Plato. *Republic*.

Playfair, J. 1802. *Illustrations of the Huttonian Theory of the Earth*. London: Cadell and Davies; Edinburgh: William Creech.

Priest, G. G. 2017. 'Charles Darwin's Theory of Moral Sentiments: What Darwin's Ethics Really Owes to Adam Smith.' *Journal of the History of Ideas* 78: 571–593.

Provine, W. 1986. *Sewall Wright and Evolutionary Biology*. Chicago: University of Chicago Press.

Quintilian. Ca 95CE. *Institutio oratoria*. Trans. and ed. H. E. Butler. Cambridge, MA: Harvard University Press. 1985.

Radick, Gregory. 2009. 'Is the Theory of Natural Selection Independent of its History?'. In Jonathan Hodge and Gregory Radick, eds., *The Cambridge Companion to Darwin*, 2nd edition. Cambridge: Cambridge University Press, pp. 147–172.

2016. 'Presidential Address: Experimenting with the Scientific Past'. *British Journal for the History of Science*. 49: 153–172.

2017. 'The Argument from Science'. In K. Almqvist and I. Thomas, eds., *Sapere Aude: The Future of the Humanities in British Universities*. Stockholm: Ax:son Johnson Foundation, pp. 185–190.

2018. 'How and Why Darwin Got Emotional About Race'. In Efram Sera-Shriar, ed., *Historicizing Humans: Deep Time, Evolution, and Race in Nineteenth-Century British Sciences*. Pittsburgh: University of Pittsburgh Press, pp. 138–171.

2019. 'So Many Free Lunches [Review of Milo 2019]'. *Times Literary Supplement* 15 Nov.: 36.

Reid, Thomas. 1785. *Essays on the Intellectual Powers of Man*.

Reznick, D. 2010. *The Origin Then and Now: An Interpretive Guide to the Origin of Species*. Princeton: Princeton University Press.

Richards, Evelleen. 2017. *Darwin and the Making of Sexual Selection*. Chicago: University of Chicago Press.

Richards, Richard A. 1997. 'Darwin and the Inefficacy of Artificial Selection'. *Studies in History and Philosophy of Science* 28: 75–97.

1998. 'Darwin, Domestic Breeding and Artificial Selection'. *Endeavour* 22: 106–109.

2005. 'Is Domestic Breeding Evidence for (or Against) Darwinian Evolution?' In Peter Achinstein, ed., *Scientific Evidence: Philosophical Theories and Applications*. Baltimore: Johns Hopkins University Press, pp. 207–236.

2014. 'Darwin's Experimentalism'. *Endeavour* 38: 235–245.

Richards, Robert J. 2002. *The Romantic Conception of Life: Science and Philosophy in the Age of Goethe*. Chicago: University of Chicago Press.

2009. "Darwin's Theory of Natural Selection and its Moral Purpose." In Ruse and Richards 2009, pp. 47–66.

Richards, Robert J. and Ruse, Michael. 2016. *Debating Darwin*. Chicago: University of Chicago Press.

Ruse, Michael and Richards, Robert J., eds. 2009. *The Cambridge Companion to the 'Origin of Species'*. Cambridge: Cambridge University Press.

Shakespeare, William. *Antony and Cleopatra*.

Shapiro, L. and Sober, Elliott. 2007. 'Epiphenomenalism – The Do's and Don't's.' In G. Wolters and P. Machamer, eds., *Studies in Causality: Historical and Contemporary*. Pittsburgh:. University of Pittsburgh Press.

Shaw, Prue. 2014. *Reading Dante from Here to Eternity*. New York: London.

Skipper, Robert. 2002. 'The Persistence of the R. A. Fisher–Sewall Wright Controversy'. *Biology and Philosophy* 17: 341–367.

Smith, Adam. 1759. *The Theory of Moral Sentiments*. London: Andrew Millar; Edinburgh: Alexander Kincaid and J. Bell.

 1776. *An Inquiry into the Nature and Causes of the Wealth of Nations*. London: W. Strahan and T. Cadell.

Sober, Elliot. 1993. *The Nature of Selection: Evolutionary Theory in Philosophical Focus*. Chicago: University of Chicago Press. First published in 1984.

 2003. 'The Design Argument'. In Neil A. Manson (ed.), *God and Design: The Telelogical Argument and Modern Science*. London: Routledge, pp. 25–53.

 2010. *Did Darwin Write the Origin Backwards? Philosophical Essays on Darwin's Theory*. Buffalo, New York: Prometheus.

Sterrett, Susan G. 2002. 'Darwin's Analogy between Artificial and Natural Selection: How Does It Go?'. *Studies in History and Philosophy of Biological and Biomedical Sciences* 33: 151–168.

Sullivan-Clarke, Andrea. 2013. 'On the Causal Efficacy of Natural Selection: A Response to Richards' Critique of the Standard Interpretation'. *Studies in History and Philosophy of Biological and Biomedical Sciences* 44: 745–755.

Theunissen, Bert. 2013. 'The Analogy Between Artificial and Natural Selection.' In Michael Ruse, ed., *The Cambridge Encyclopedia of Darwin and Evolutionary Thought*. Cambridge: Cambridge University Press, pp 88–94.

Wallace, Alfred Russel. 1858. 'On the Tendency of Varieties to Depart Indefinitely From the Original Type.' *Journal of the Proceedings of the Linnean Society of London. Zoology* 3: 45–62. Reprinted in Gavin de Beer, ed., *Evolution by Natural Selection: Darwin and Wallace*. Cambridge: Cambridge University Press, 1958.

Wallace, A.R. 1864. 'The Origin of Human Races and the Antiquity of Man Deduced from the Theory of "Natural Selection"'. *Journal of the Anthropological Society of London* 2: clviii–clxx.

 1889. *Darwinism: An Exposition of the Theory of Natural Selection, with Some of Its Applications*. London: Macmillan.

 1890. 'Human Selection.' *Fortnightly Review* 48 (n.s.): 325–337.

Walsh, Denis, Ariew, André and Matthen, Mohan. 2017. 'Four Pillars of Statisticalism.' *Philosophy, Theory and Practice in Biology* 9: 1–19.

Waters, C. Kenneth. 1986. 'Taking Analogical Inference Seriously: Darwin's Argument from Artificial Selection'. *Proceedings of the Biennial Meeting of the Philosophy of Science Association* 1: 502–513.

Weitzenfeld, Julian. 1984. 'Valid Reasoning by Analogy'. *Philosophy of Science* 51: 137–149.

West, J. B. 2001. 'Snorkel Breathing in the Elephant Explains the Unique Anatomy of its Pleura'. *Respiratory Physiology & Neurobiology* 126:1–8.

Wetherbee, W. and Alexander, J. 2018. 'Dante Alighieri.' In the *Stanford Encyclopedia of Philosophy*, online.

Whately, Richard. 1822. *On the Use and Abuse of Party Feeling in Matters of Religion*. Bampton Lectures. Oxford: Oxford University Press.

 1826. *Elements of Logic, Comprising the Substance of the Article in the Encyclopaedia Metropolitana: With Additions*. London: Mawman.

1828. *Elements of Rhetoric*. London: John Murray; Oxford: J. Parker.

White, Roger M. 1982. 'Notes on Analogical Predication, and Speaking about God'. In B. Hebblethwaite and S. Sutherland, eds., *The Philosophical Frontiers of Christian Theology*. Cambridge: Cambridge University Press, pp. 197–226.

1996. *The Structure of Metaphor: The Way the Language of Metaphor Works*. Oxford: Blackwell.

2001. 'Literal Meaning and "Figurative Meaning"'. *Theoria* 67: 24–59.

2010. *Talking about God: The Concept of Analogy and the Problem of Religious Language*. Farnham: Ashgate.

2015. 'Analogy, Metaphor and Literal Language.' In G. Oppy, ed., *The Routledge Handbook of Contemporary Philosophy of Religion*. London: Routledge, ch. 17.

Winsor, Mary Pickard. 2013. 'Darwin and Taxonomy'. In Michael Ruse, ed., *The Cambridge Encyclopedia of Darwin and Evolutionary Thought*. Cambridge: Cambridge University Press, pp. 72–79.

Wright, Sewall. 1931. 'Evolution in Mendelian Populations'. *Genetics* 16: 97–159. Reprinted in W. Provine, ed., *Sewall Wright: Evolution. Selected Papers*. Chicago: University of Chicago Press, 1986.

1932. 'The Roles of Mutation, Inbreeding, Cross Breeding and Selection in Evolution'. *Proceedings of the Sixth International Congress of Genetics* 1, pp 356–366. Reprinted in W. Provine, ed., *Sewall Wright: Evolution. Selected Papers*. Chicago: University of Chicago Press, 1986.

1978. 'The Relation of Livestock Breeding to Theories of Evolution.' *Journal of Animal Science* 46: 1192–1200. Reprinted in W. Provine, ed., *Sewall Wright: Evolution. Selected Papers*. Chicago: University of Chicago Press, 1986.

Young, Robert M. 1971. 'Darwin's Metaphor: Does Nature Select?'. In his *Darwin's Metaphor: Nature's Place in Victorian Culture*. Cambridge: Cambridge University Press, 1985, pp. 79–124 (with new postscript on pp. 124–5). First published in the *Monist* 55: 442–503.

Index

Printed in the United States
by Baker & Taylor Publisher Services

Printed in the United States
by Baker & Taylor Publisher Services